DISTIL
FLAVOU
1989.

S0-BKB-941

10000648843904
HEAL

Piggott/Paterson

Distilled Beverage Flavour

Ellis Horwood Series in Food Science and Technology

Series Editors: I. D. Morton, formerly Head of Department of Food Science, King's College (KQC), Kensington Campus, Campden Hill Road, London; R. Scott, formerly University of Reading; D. H. Watson, Ministry of Agriculture, Fisheries and Food; M. Lewis, Department of Food Science, University of Reading.

Published Titles

parat Dictionary of Food and Nutrition J. Adrian, G. Legrand and R. Frangne

Microwave Processing and Engineering R. V. Decareau and R. A. Peterson

Food Processing Technology: Theory, Applications, Equipment P. Fellows

Fundamentals of Food Chemistry W. Heimann

Nitrosamines: Toxicology and Microbiology M. J. Hill (Ed.)

Sensory Evaluation of Food: Theory and Practice G. Jellinek

Hygienic Design and Operation of Food Plant R. Jowitt (Ed.)

Services, Heating and Equipment for Home Economists D. Kirk and A. Milson

Sustainable Food Systems D. Knorr (Ed.)

Physical Properties of Foods and Food Processing Systems M. J. Lewis

Technology of Biscuits, Crackers and Cookies D. J. R. Manley

Principles of Design and Operation of Catering Equipment A. Milson and D. Kirk

Cereals in a European Context: First European Conference on Food Science and Technology I. D. Morton (Ed.)

Animal By-Product Processing H. W. Ockerman and C. L. Hansen

The Role of Fats in Human Nutrition F. B. Padley and J. Podmore (Ed.)

Flavour of Distilled Beverages: Origin and Development J. R. Piggott

Distilled Beverage Flavour: Recent Developments J. R. Piggott, A. Paterson (Eds.)

Advanced Sugar Chemistry: Principles of Sugar Stereochemistry R. S. Shallenberger

Egg and Poultry-Meat Products W. J. Stadelman, V. M. Olson, G. A. Shemwell and S. Pasch

Frying of Food: Principles, Changes, New Approaches G. Varela, A. E. Bender and I. D. Morton (Eds.)

Natural Toxicants in Food: Progress and Prospects D. H. Watson (Ed.)

Sensory Quality in Foods and Beverages A. A. Williams and R. K. Atkin (Eds.)

© Ellis Horwood Ltd., Chichester (England), 1989

Distribution:

Great Britain and Ireland: VCH Publishers (UK) Ltd., 8 Wellington Court, Wellington Street, Cambridge CB1 1HW (Great Britain)

USA and Canada: VCH Publishers, Suite 909, 220 East 23rd Street, New York, NY 10010-4606 (USA)

Switzerland: VCH Verlags-AG, P. O. Box, CH-4020 Basel (Switzerland)

All other countries: VCH Verlagsgesellschaft, P. O. Box 1260/1280, D-6940 Weinheim (Federal Republic of Germany)

ISBN 3-527-26887-1 (VCH Verlagsgesellschaft) ISBN 0-89573-819-8 (VCH Publishers)

Distilled Beverage Flavour

Recent Developments

Edited by
J. R. Piggott and A. Paterson

John R. Piggott
Lecturer in Food Science
Department of Bioscience & Biotechnology
University of Strathclyde
131 Albion Street
Glasgow G1 1SD

A. Paterson
Lecturer in Food Science
Department of Bioscience & Biotechnology
University of Strathclyde
131 Albion Street
Glasgow G1 1SD

Deutsche Bibliothek Cataloguing-in-Publication Data
Distilled beverage flavour : recent developments / ed. by
J. R. Piggott and A. Paterson. – Cambridge ; New York, NY ;
Basel (Switzerland) ; Weinheim : VCH ; Chichester, England :
Horwood, 1989.
(Ellis Horwood series in food science and technology)
ISBN 3-527-26887-1 (VCH, Weinheim) Pp.
ISBN 0-89573-819-8 (VCH, Cambridge ...) Gb.
NE: Piggott, John Raymond [Hrsg.]

ISSN 0930-3332

British Library Cataloguing in Publication Data
Distilled beverage flavour.
1. Alcoholic drinks. Flavours. Chemical analysis
I. Piggott, J. R. (John Raymond) II. Paterson, A.
663'.10287

Library of Congress Card No. 88–35234

Published jointly in 1989 by
Ellis Horwood Ltd., Chichester (England), VCH Verlagsgesellschaft mbH, Weinheim (Federal Republic of Germany) and VCH Publishers, New York, NY (USA)

All Rights Reserved. No part of this publication may be reproduced, stored in a retrieval system, or transmitted, in any form or by any means, electronic, mechanical, photocopying, recording or otherwise, without the permission of Ellis Horwood Limited, Market Cross House, Cooper Street, Chichester, West Sussex, England.
Registered names, trademarks, etc. used in this book, even when not specifically marked as such, are not to be considered unprotected by law.

Printed and Bound in Great Britain by Hartnolls Limited, Bodmin, Cornwall

Table of Contents

Preface

This book represents the proceedings of an international symposium held at Stirling University, Scotland, between 7th and 10th June 1988, by the Sensory Panel of the Food Group of the Society of Chemical Industry. The symposium was held to review advances in understanding of distilled beverage flavour made since the first symposium in 1983. At the first meeting, the current status of knowledge was reviewed for the main groups of products; at this meeting it was clear that progress has been made in the ability to control flavours of distilled beverages in some areas, but that many problems remain to be solved. Particular problems identified were the absence of any satisfactory means of linking sensory and physico-chemical data for complex flavours, though in some cases this can be done; incomplete understanding of the maturation process in traditional matured beverages; and the problem of authentication of beverages. Many opportunities were also identified.

The chapters here are arranged in four sections. The first section is devoted to methods of analysis, both sensory and instrumental, of distilled beverage flavours and sensory properties. The second section contains contributions discussing traditional products of the distilled beverage industry, while the third section covers the opportunities and problems of new products. Finally, some new methods and processes for production are described. The book is completed by a final chapter summarising progress since 1983.

We are pleased to acknowledge the assistance given by many people in arranging this symposium: the staff of Stirling University for their help and hospitality; the staff of the Society of Chemical Industry for their patience; the hosts of the social events for making the symposium enjoyable as well as instructive; the contributors for producing typescripts promptly; and the publishers.

J.R. Piggott and A. Paterson

1

Current issues in flavour research

Henk Maarse and Frans van den Berg
TNO-CIVO Food Analysis Institutes,
Zeist, The Netherlands

1. INTRODUCTION

Application of chromatography, particularly in combination with mass spectrometry, has led to the identification of a large number of constituents in distilled beverages [1,2,3]. Despite these impressive results more attention should be given to sensory analysis to help the flavour analyst in selecting those compounds that are important for the flavour of a product.

In Section 2 we will discuss this topic by describing sensory techniques that are being used in combination with chromatographic analysis.

New developments in chromatographic techniques in combination with mass spectrometry and Fourrier Transform infra-red spectrometry have recently been extensively reviewed [4] and will not be discussed in this paper.

The maturation flavour is a very important part of the overall flavour of alcoholic beverages such as cognac and whisky. New developments in this field will be reviewed in Section 3.

For some people adulteration of alcoholic beverages is an attractive way of making money. Although many administrative measures have been taken, these do not fully prevent frauds. Thus, modern techniques are being proposed and tested in many countries. In Section 4 the possible uses of these techniques in the determination of the origin of wine distillates are shown.

2. COORDINATION OF SENSORY AND INSTRUMENTAL ANALYSES

2.1 Multidimensional gas chromatography

An elegant way to improve the separation power of a GC system is to use two columns in series, a pre-column and an analytical column. Interesting fractions from the first column can be introduced on-line into the second column to be further separated into individual components.

Good examples of multidimensional GC (MDGC) in flavour research are the studies of Nitz et al.[5,6] who used a GC with two ovens, which were independently temperature-programmable (Siemens Model Sichromat 2). A scheme of the total system is depicted in Figure 1-a. A detailed description of this system, based on the methods introduced by Deans [7] and further developed by Schomburg et al. [8,9], can be found in the original literature [5,6]. Below the procedure is described and the possibilities of the system are shown.

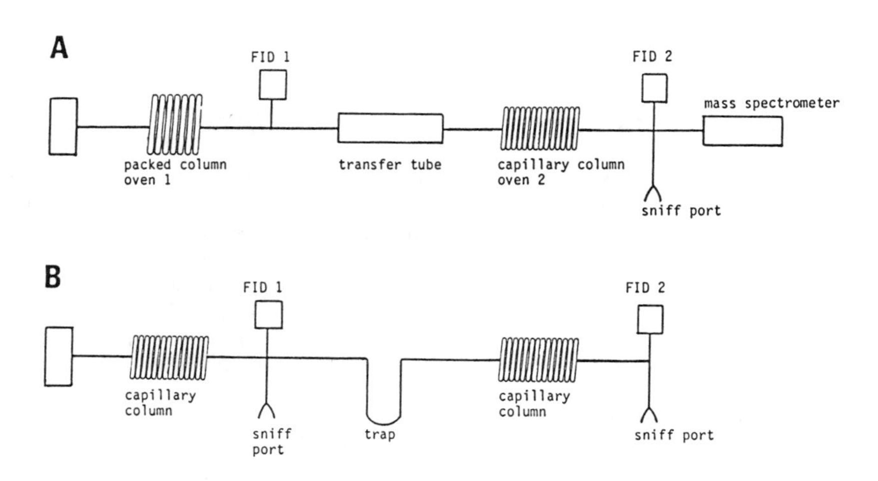

Figure 1 Two-column switching systems for multidimensional GC
a. Schematic drawing of the Siemens system (Model Sichromat 2) for total transfer from packed to capillary columns with intermediate trapping and a device for parallel MS/sniffing or MS/FID registration.
b. Schematic drawing of the Chrompack Music system for total transfer from pre-columns to analytical columns with intermediate trapping and a device for parallel sniffing/FID registration or trapping/FID registration.

Depending on the problem the pre-column may be a packed column, a medium-resolution wide-bore capillary column or a high-resolution capillary column of similar dimensions but with a polarity differing from the analytical column.

In the examples described in the literature the first column was a packed column and the second one a capillary column. Selected fractions from the first column were trapped on a cooled transfer tube. The trapped compounds were re-injected into the second column by heating the tube. Compounds separated on this column can be directly monitored with a FID detector or the effluent can be split for simultaneous MS registration. Furthermore, parallel sniffing and MS detection can be performed.

Nitz and Julich [5] used this system to trace and identify the volatile components characteristic of the flavour of cooked cauliflower. Preliminary analysis by GC-MS was not successful: the mass spectra of relevant compounds were of low purity and could not be interpreted. Therefore, 4 μl of the concentrate were injected on to the packed column of the MDGC system, the fractions between 19.4 and 20.6 min and between 23 and 26 min were trapped and re-injected on to the second column (Figure 2-a).

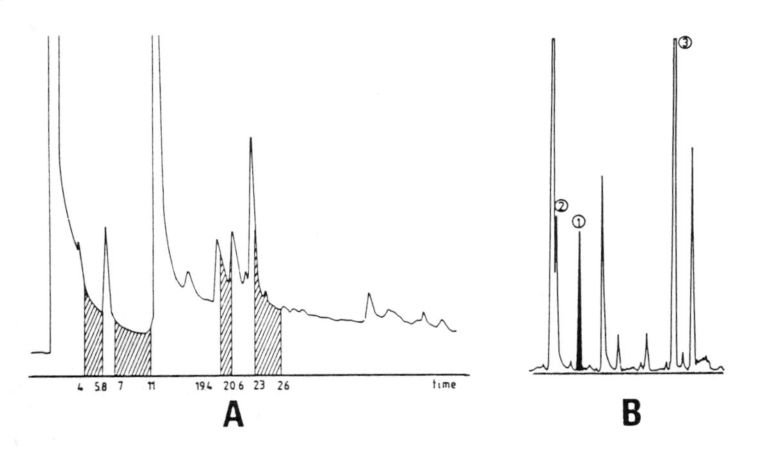

Figure 2.a. Chromatogram of cooked cauliflower extract obtained on packed pre-column; cutting areas are indicated [5].
b. Total ion chromatogram of definite cuts of cauliflower extracts; fraction 19.4-20.6 and 23-26 minutes [5].

Simultaneous sniffing during MS acquisition showed that the component marked (1) in Figure 2-b was the one with a cooked cauliflower odour. A clean mass spectrum was now obtained. In our institute we have chosen a cheaper solution, the two-dimensional switching system called Music (Multiple Switching Intelligent Controller) developed by Chrompack in the Netherlands (10).

The set-up of this system is depicted in Figure 1-b: the pre-column is a 10 m * 0.53 mm (i.d.) CP Sil5 fused silica column, analytical column a 25 m * 0.25mm (i.d.) CP Wax 52CB fused silica column.

We adapted the system by introducing a split, either between the pre-column and the detector or between the analytical column and the detector.

By this 1:10 split 10% of the effluent was led to the detector and 90% to a sniff port.

We used this system for the analysis of a commercial fruit flavour. Three fractions were first selected on the apolar column possessing a flavour characteristic of this particular fruit.

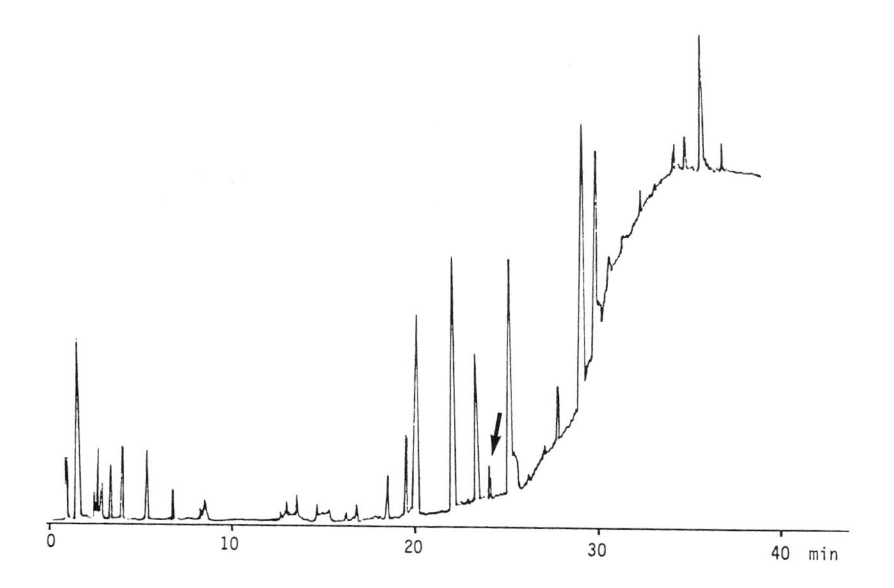

Figure 3. Chromatogram of a fraction with a characteristic flavour: the arrow indicates the place in the chromatogram where a catty odour was observed after further separation of a selected column fraction.

In a second run these three fractions were trapped on a piece of de-activated fused silica capillary. These fractions were further separated on the second, polar column. By sniffing, the peaks responsible for the characteristic flavour were selected. In a final run the fractions were trapped and re-injected into a GC-MS system and the selected peaks were identified. The device for intermediate trapping was also used for trapping the fractions for GC-MS analysis. In one of the fractions an important flavour was described as Buchu-like (Figure 3).

This small peak was identified as 8-mercapto-p-menthan-3-one; this compound has been identified in Buchu oil [11], it has a very low odour threshold value and its odour is described as catty or ribes-like. This illustrates that our system is a very powerful tool in identifying flavour compounds.

We also use the MUSIC system for the quantitative analysis of ethyl carbamate in alcoholic beverages [12] and of carbitol in flavour concentrates [13].

2.2 Determination of odour threshold values

Odour and flavour threshold values are important properties of odorous volatile compounds. Knowledge of these values is indespensible in order:

- to get an idea of the importance of compounds for the flavour of a product.
- to find out whether the instrumental detection limit is sufficiently low, preferably at least as low as the threshold value. This plays a part only for compounds contributing negatively to the flavour of a product.
- to estimate maximum allowable concentrations of a compounds causing off-odours and/or off-flavours in foods, raw materials and related products (e.g. packaging materials [14]).

The outcome of threshold value determinations is strongly dependent on the experimental conditions. Interlaboratory results can easily differ by a factor of 10 and even more. Sensory experiments with compounds having very low threshold certainly require careful consideration of all relevant aspects [14].

In our laboratory odour detection threshold values of several chlorophenols and chloroanisoles have been determined. The results of this study have been reported elsewhere [14]. Only the description of one of the techniques used in that study is relevant to the scope of this paper.

For the determination of the odour threshold values in air an olfactometer was used. This is an apparatus capable of producing airflows with a wide range of concentrations of volatile compounds. This airflow is presented, together with two odourless airflows, to a panel of 8 members. After smelling they are asked to indicate which of the three is the odorous one.

To check the purity of the compounds, standard solutions were first injected splitless on to a GC column. The end of the column was connected to a splitter which led part of the effluent to a FID detector and the other part directly out of the oven. During elution a 45-litre Teflon bag was connected to the splitter outlet. The concentration of the pure compound in the bag was calculated from the concentration of the standard solution, the injection volume and the split ratio at the end of the column. In this way, odour threshold values of chloroanisoles have been determined [14].

2.3 Selection of compounds contributing significantly to flavours

2.3.1 *Gas Chromatography*

All foods and beverages contain hundreds of volatile compounds and it is therefore essential to select those constituents that contribute significantly to the flavour of a product.

Roche and Thomas [15] suggested the use of 'aroma value', being the ratio of the concentration of a flavour compound in a product to its odour threshold value. Guadagni et al. [16] called the same factor 'odour unit' and Mulders [17] 'odour value'. In this paper the term 'odour unit' is used. Although this approach has its limitations and has been criticised [e.g. 18] its usefulness is beyond all doubt.

Odour units can be calculated only if the concentration and the threshold value of compounds are known. This means that major flavour compounds are selected after identification of a large number of constituents. This is very laborious and costly and does not guarantee that all important flavour compounds are taken into consideration.

As described in Section 2 the sniff technique is being used by many investigators to provide odour information on compounds separated by GC. Recording the intensity of the perceived odour gives additional information on the contribution of compounds to the flavour of a product [e.g. 19]. Recently two new methods for determining these volatiles in food that are significant of flavour have been presented [20 - 22].

Acree et al. applied their method to volatiles in apples [23]. Their procedure is also based on the 'odour unit' concept, but they determine relative odour detection thresholds of components in a flavour extract using the GC-sniff technique. They calculated 'Charm Values' on the basis of sensory responses maintained during GC effluent sniffing of dilutions of the original extract. Charm values are directly proportional to odour units.

Grosch et al. [21,22] used a similar but simpler method. An aroma extract is stepwise diluted with a solvent until no more odorous compounds are observed in the GC effluent. The highest dilution at which a substance is still smelled is its flavour dilution (FD) factor. By definition the FD factor of an undiluted sample is 1. An FD factor of 20 means that the concentration of the odorous compound in the aroma extract was 20 times its odour threshold value as perceived by the GC effluent sniffing. FD factors are proportional to odour units.

This aroma extract dilution analysis was applied by Schieberle and Grosch [22] to wheat and rye bread crusts. Part of the results are given in Figure 4 showing the FD factor versus the retention index in an 'aromagram'. This name is somewaht confusing: other authors understand aromagrams to mean chromatograms, indicating also the flavour descriptions of the compounds observed at a sniff port. We therefore suggest the name 'flavour dilution chromatogram'.

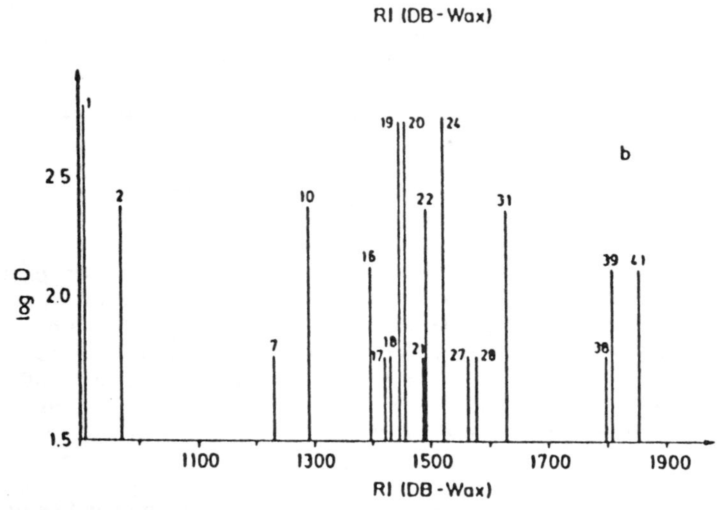

Figure 4. Flavour dilution chromatogram of the aroma extract of rye bread crust. Peaks numbered as follows:

1. 3-methylbutanal 2. diacetyl

7. 4-cis-heptenal 10. 1-octen-3-one

16.a.trimethylpyrazine 16.b.2-methyl-3-ethylpyrazine

17. 2-tr-octenal 18. 2,5-dimethyl-3-ethylpyrazine

19. unknown 20.a. furan-2-al

20.b. 2,6-dimethyl-3-ethylpyrazine 21 & 22. unknown

24. 2-tr-nonenal 27. 5-methyl-2-furaldehyde

28. 2-tr-6-cis-nonadienal 31. phenylacetaldehyde

28. 2-tr-4-tr-decadienal 39 & 41. unknown

The results clearly show that some compounds that contribute to the flavour of rye bread crust are still unknown.

2.3.2 High-pressure liquid chromatography (HPLC)

In the analysis of distilled alcoholic beverages HPLC is routinely used for the determination of acids, phenols, phenolic acids and aldehydes, and sugars. Also methods for the determination of aliphatic carbonyl compounds, as their dinitrophenylhydrazone derivitives, have been worked out [24].

Apart from these quantitative HPLC analyses, identification of unknown high-boiling or non-volatile flavour compounds is possible with HPLC-MS combinations [25] which are now commercially available.

In the literature no methods have been described that enable the selection of important flavour compounds in the HPLC eluent.

The reasons are obvious:

o a "taste port" cannot be used because tasting takes more time than smelling.

o the composition of most HLPC eluents is not suitable for tasting. They either are toxic or they fully disturb the observation of flavour compounds in a fraction.

Therefore an eluent should be chosen that :

- is not toxic
- has no or only a very weak flavour and
- still enables sufficient separation.

With such a system we have fractioned a wood extract. It was directly introduced and eluted with a gradient of ethanol and a phosphoric acid-water mixture from an RP-18 column. Aromatic aldehydes were quantified using a diode-array detector and the eluent was collected in fractions. These were brought to pH4 and subsequently submitted to sensory analysis. Some results are described in the next section.

Table 1. Quantified constituents in brandy and oak wood extracts

Volatiles	
phenol	26,27
o-,m-,p-cresol	26,27
guaiacol	26,27
ethylguaiacol	26,27
eugenol	26,27
-methyl- -octalactone	31

Coumarines	
aesculetine	28
umbelliferone	28
scopoletine	28
methylumbelliferone	28

Sugars	
glucose	34
mannose	34
xylose	34
arabinose	34
galactose	34
rhamnose	34
protoquersitol	34

Aromatics	
ligno-complex	30,31,32,33
lignin	30,32
vanillin	26,28,29,31-33
syringaldehyde	26,28,29,31-33
coniferaldehyde	28,31-33
protocatechualdehyde	29
sinapaldehyde	28,31-33
vanillic acid	26,29,31-33
syringic acid	29,31-33
ferulic acid	31,32
p-coumaric acid	31,32
cinnamic acid	31,32
protocatechuic acid	26,29,31
-hydroxybenzoic acid	26,28,31
gallic acid	26,31
benzoic acid	32

non-volatiles	
tannins	30,31
5-hydroxymaltol	26

3. MATURATION FLAVOUR OF BRANDIES

Distilled beverages such as cognac and armagnac are matured in oak barrels. This traditional process has a strong effect on the colour, taste and odour of these products.

During the past 10 years many investigations have been reported, providing essential information on the chemical changes taking place during the maturation period. Table 1 overviews the compounds quantified in brandies and wood extracts.

Lignin degradation products such as vanillin and vanillin-like aldehydes and acids are regarded as important contributors to the maturation flavour.

Maga [35] has determined the taste threshold values of vanillin , syringaldehyde, sinapaldehyde, ferulic acid, vanillic acid, syringic acid and sinapic acid in 40% (v/v) alcoholic solutions. The threshold values of the individual compounds, apart from vanillin, are higher than the levels found in brandy [29,31,32]. In the same investigation, however, a strong synergistic effect was noticed resulting in much lower thresholds. Taking this synergistic effect into account, these compounds most likely contribute to the woody flavour. Puech has studied their occurrence in cognac [31], armagnac [32] and oak wood extracts used in aging of brandy [33].

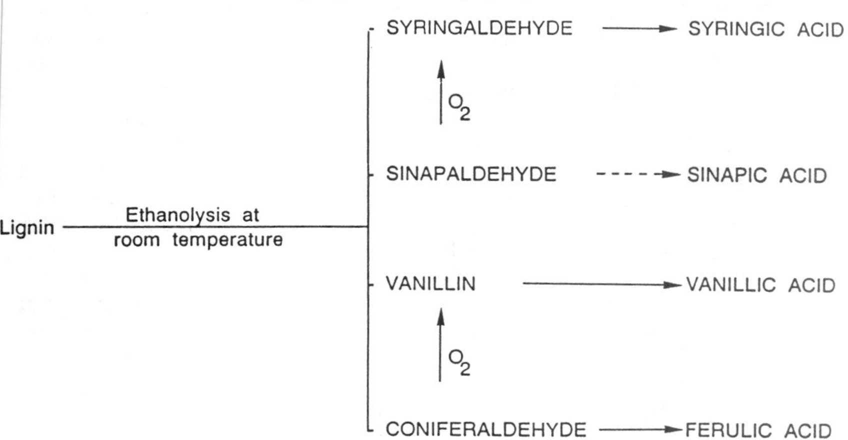

Figure 5. Model of degradation of lignin during brandy maturation, according to Puech [31,32].

Puech has proposed a model for the degradation of lignin (Figure 5). In this model some degradation pathways are presented indicating the influence of oxygen. We have studied the influence of oxygen on the degradation of lignin during the extraction of Limousine oak wood with alcoholic solutions having different oxygen concentrations. We noticed that the presence of oxygen has a strong effect on the degradation of lignin. An increasing oxygen content results in higher concentrations of vanillin, syringaldehyde, coniferaldehyde, vanillic acid and syringic acid. Extracts prepared with little oxygen contained more sinapaldehyde. The presence of oxygen in the extraction fluid did not significantly affect the concentrations of sinapic and ferulic acid. Apart from the lignin degradation products an increase of eugenol and furfural was noticed in oxygen-containing extracts.

The extracts were also compared by sensory analysis. Despite the higher contents of phenolic aldehydes and acids in the extracts with oxygen-containing ethanol-water mixtures extracts prepared with little oxygen were found to have a more harmonic, cognac-like and less astringent flavour. Sensory comparison of HPLC fractions (see Section 2.3.2) of different types of wood extracts showed also large differences outside the part of the chromatogram where the aromatic aldehydes and acids were eluted. The compounds in these fractions are yet to be identified.

It may be concluded that oxygen has a strong influence on the degradation of lignin. The sensory results also indicate that constituents other than the well-known lignin degradation products contribute to the maturation flavour. Concentrations of volatile compounds such as β-methyl-γ-octalactone (oak lactone) and phenols increase during maturation of a brandy. The ranges of concentration of phenols have been determined by Lethonen [26,27]. Phenol, p-ethylphenol, p-ethylguaiacol and eugenol were found to be present in brandy at concentrations between 0.01 and 0.3 mg/l. Swan et al. [36] have determined the odour threshold values of these components in whisky. Comparing these data with the analytical results of Lehtonen one may conclude that apart all the phenols except phenol contribute to maturation flavours.

Oak lactone is a very interesting compound. It has a clear woody odour at low concentrations. Puech [31] has quantified this compound in many cognacs of different ages and has found concentrations varying between 0.4 and 2.9 mg/l depending on the age of the product. Sharp [37] has studied the sensory properties of oak lactone in 40% (v/v) ethanol and has reported that it has a woody flavour at low concentrations and a coconut-like odour prevailing at high concentrations. The concentrations of oak lactone in cognac as reported by Puech are similar to the values found by Sharp to cause a woody flavour.

Another group of compounds extracted from wood are the coumarines. Salagoity-Auguste et al. [28] have analysed the content of aesculetine, umbelliferone, scopoletine and methyl-umbelliferone in armagnac and calvados. They noticed that aesculetine and scopoletine were extracted in relative high concentrations (1.9 and 0.3 mg/l respectively). The sensory properties of these compounds are unknown. Carbohydrates such as glucose, proto-quersitol, arabinose, xylose, galactose, rhamnose and mannose are extracted from wood as degradation products of hydrolysed hemicelluloses [34]. The effect of these compounds on the flavour of brandy is unknown.

Many different compounds contribute to the maturation flavour. To evaluate the contribution of these compounds to the characteristic woody flavour we have carried out sensory tests with synthetic mixtures of phenols, tannins, oak lactone and aromatic aldehydes and acids in concentrations similar to those found in cognac. From the results of these tests we conclude that compared to a naturally-aged wine distillate, both taste and odour of the wine distillate with the synthetic maturation mixture lacks some typical woody characteristics.

During the past 10 years much progress has been made in broadening the insight into the compounds responsible for the maturation flavour of brandy.

Further studies applying combined sensory and analytical techniques such as GC-sniffing/GC-MS, sensory evaluation of HPLC fractions combined with HPLC-MS, and the use of descriptive panels, are needed to unravel this complex flavour.

More is known about the flavour of whisky. The reason might well be that sophisticated sensory techniques in combination with instrumental analysis are being applied more frequently in the United Kingdom than in some countries in continental Europe.

4. AUTHENTICITY OF DISTILLED BEVERAGES

Recently adulteration of wines in Austria with diethylene glycol and in Italy with methanol has caused wine scandals [38]. Producers of distilled beverages have been afraid that they will fall victim to further scandals. Protection, however, against fraudulent actions is very difficult. One way is to thoroughly determine the composition of a product and to show that no foreign compounds are present. Furthermore, natural compounds should be present in concentrations within the ranges of concentrations of compounds in the relevant distilled alcoholic beverage. These ranges can be found in one of our publications [39]. We have used this method to analyse wine distillates, used in the production of brandies, produced in France, Spain and Italy. Two gas chromatograms, one of the low-boiling compounds and one of the less volatile compounds, provided quantative data on 48 of the constituents.

These chromatograms showed that no foreign compounds were present and that the concentration of all constituents were within the normal range. We also found that multivariate statistical analysis of these data can be used to check the origin of the distillates. In Figure 6 it can be seen, that a clear differentiation between wines produced in different areas was found. Although these results should be confirmed by analysing more samples, the first results are promising.

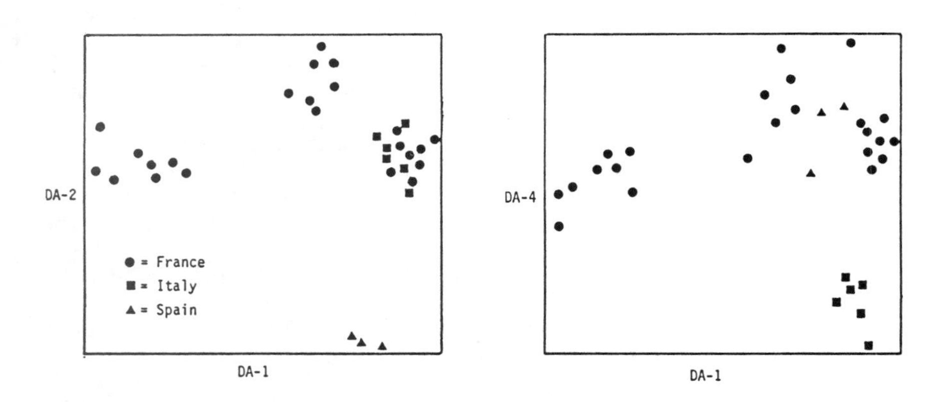

Figure 6. Scatter plot of two discriminant functions.

Cantagrel [40] used the same approach to differentiate between cognac and brandies.

5. CONCLUSIONS

- In instrumental analysis of distilled alcoholic beverages more attention should be given to the contribution of compounds to the flavour of a product.

- Multidimensional gas chromatographs are now commercially available; their use in combination with a sniff port is very valuable in flavour research.

- The flavour dilution technique following the odour unit concept, though theoretically not fully justified, is very useful for the selection of compounds contributing significantly to flavour.

- Although maturation has a pronounced influence on the flavour of aged beverages most investigators have so far studied it by instrumental analytical techniques only. Sensory analysis can broaden the insight into this process.

- The origin of wine distillates can be determined on the basis of multidimensional statistical analysis of GC data.

6. LITERATURE

1. Maarse, H. & Visscher, C.A. (1982 - 1987)
Volatile compounds in food.
TNO-CIVO Food Analysis Institute, Zeist, The Netherlands.
2. Nykanen, L. & Suomalainen, H. (1983)
Aroma of beer, wine and distilled alcoholic beverages.
Akademie Verlag, Berlin.
3. Heide, R. ter (1986)
The flavour of distilled beverages. pp. 239 - 336.
In : Food Flavours. Part B. The flavour of beverages.
Morton, I.D. and Macleod, A.J., eds.,
Elsevier, Amsterdam.
4. Schreier, P. (1988)
On-line coupled HRGC techniques for flavour analysis.
Proc. 2nd Wartburg Aroma Symposium, November 1987.
Rothe, M. ed.. Akademie Verlag, in press.
5. Nitz, S. & Julich, E. (1984)
Concentration and GC-MS analysis of trace volatiles by sorption-desorption techniques. pp. 151 - 170.
In : Analysis of volatiles.
Schreier, P. ed.. W. de Gruyter, Berlin.
6. Nitz, S., Kollmansberger, H. & Drawert, F. (1988)
Analysis of flavours by means of combined cryogenic headspace enrichment and multidimensional gas chromatography.
In : Bioflavour '87. pp. 123 - 136.
Schreier, P, ed.. W. de Gruyter, Berlin.
7. Deans, D.R. (1968)
A new technique for heart cutting in gas chromatography.
Chromatographia **1**, 18 - 22.
8. Schomburg, G., Dielmann, R., Husmann, H. & Weeke, F. (1976)
Aspects of double-column gas chromatography with glass capillaries involving intermediate trapping.
J. Chromatogr. **122**, 55 - 72.
10. Chrompack News Special 1985, 85-3
11. Nijssen, L.M. & Maarse. H. (1986)
Volatile compounds in blackcurrant products.
Flavour and Fragrances J., **1**, 143 - 148.
12. Ingen, R.H.M. van, Nijssen, L.M., Berg, F. van den & Maarse, H. (1987)
Determination of ethyl carbamate in alcoholic beverages by two-dimensional gas chromatography.
J. High Resol. Chrom. & Chrom. Commun. **10**, 151 - 152.
13. Ingen, R.H.M. van & Nijssen, L.M. (1988)
Determination of diethyleneglycolmonoethylether (carbitol) in flavours by means of two-dimensional chromatography.
J. High Resol. Chrom. & Chrom. Commun. Submitted.
14. Maarse, H., Nijssen, L.M. & Angelino, S.A.G.F. (1987)
Halogenated phenols and chloroanisoles: occurrence, formation and prevention.
Proc. 2nd Wartburg Aroma Symposium, November 1987.
Rothe, M. ed.. Akademie Verlag, in press.

15. Rothe, M. & Thomas, B. (1963)
Aromastoffe des brotes.
Z. Lebensm. Unters. Forsch. **119**, 302 - 310.
16. Guadagni, D.G., Buttery, R.G. & Harris, J. (1966)
Odour intensities of hop oil components.
J. Sci. Food Agric. **17**, 142 - 144.
17. Mulders, E. (1973)
The odour of white bread. IV Quantitative determination of constituents in vapour and their odour values.
Z. Lebensm. Unters. Forch. **151**, 310 - 317.
18. Frijters, J.E.R. (1979)
Some phycophysical notes on the use of odour unit number. pp. 47 - 51. In : Progress in Flavour Research.
Land, D.G and Nursten, H.E. eds., Applied Science, London.
19. Swan, J.S., Howie, D., Burtles, S.M., Williams, A.A. & Lewis, M.F. (1981).
Sensory and instrumental studies of Scotch whisky flavour.
In : The Quality of Food and Beverages. pp. 201 - 223.
Vol. 1. Chemistry and Technology.
Charalambous, G. and Inglett, G. (eds.)
Academic Press, New York, 1981.
20. Acree, T.E., Barnard, J. & Cunningham, D.G. (1984)
A procedure for the sensory analysis of gas chromatographic effluents.
Food Chem. **14**, 273 - 286.
21. Grosch, W. & Schieberle, P. (1987)
Bread flavour: qualititative and quantitive analysis.
Proc. 2nd Wartburg Aroma Symposium.
Rothe, M. ed.. Akademie Verlag, in press.
22. Schieberle, P. & Grosch, W. (1987)
Evaluation of the flavour of wheat and bread crusts by aroma extract dilution analysis.
Z. Lebensm. Unters. Forsch. **185**, 111 - 113.
23. Cunningham, D.G., Acree, T.E., Barnard, J., Butts, R.M. & Braell, P.A. (1986)
Charm analysis of apple volatiles.
Food Chem. **19**, 137 - 147.
24. Poputti, E. & Lethonen, P. (1986)
High-performance liquid chromatographic separation and diode-array spectroscopic identification of dinitrophenylhydrazone derivatives of carbonyl compounds.
J. Chromatogr. **353**, 163 - 168.
25. Maarse, H (1984)
Methods of separating and identifying flavour compounds.
Proc. Alko Symp. Flavour research of alcoholic beverages.
Nykanen, L; Lethonen, P. eds.; Foundation for Biotechnical and Industrial Fermentation Research **3**, 71 - 97.
26. Lethonen, M. & Jounela-Eriksson, P. (1983)
Volatile and non-volatile compounds in the flavour of alcoholic beverages. pp. 64 - 78.
In: Flavour of Distilled Beverages,
Piggott, J.R. ed.. Ellis Horwood Ltd, Chichester.

27. Lethonen, M. (1983)
Gas-liquid chromatographic determination of volatile phenols in matured distilled alcoholic beverages.
J. Assoc. Off. Anal. Chem. **66**, 62 - 70.
28. Auguste, M.H., Tricard, C. & Surraud, P. (1987)
Dosage simultane des aldehydes aromatiques et des coumarines par chromatographie liquide haute performance.
J. Chromatogr. **392**, 379 - 387.
29. Sontag, G. & Friedrich, O. (1988)
Bestimmung phenolischer Verbindungen in alkoholischen Getranke durch HPLC mit elektrochemischen Detektor.
Z. Lebensm. Unters. Forsch. **186**, 130 - 133.
30. Puech, J.L.; Jouret, C.; Goffinet, B.
Evolution des composes phenolic du bois de chene au cours de vieillissement de l'Armagnac. Sci. Aliments. **5**, 379 - 391.
31. Puech, J.L., Leauate, R., Clot, G. & Nomdedue, L. (1984)
Evolution de divers constituents volatiles et phenolique des eau de vie de Cognac au cours de leur vieillissement.
Sci. Aliments. **4**, 65 - 80.
32. Puech, J.L. (1981)
Extraction and evolution of lignin products in armagnac matured in oak. Am. J. Enol. Vitic. **32**, 111 - 114.
33. Puech, J.L. (1988)
Phenolic compounds in oak wood extracts used in the aging of brandies. J. Sci. Food. Agric. **42**, 165 - 172.
34. Nykanen, L., Nykanen, I., & Moring, M. (1984)
Aroma Compounds dissolved from oak chips by alcohol.
pp. 339 - 346. In : Progress in Flavour Research.
Adda, J. ed.. Elsevier, Amsterdam.
35. Maga, J.A. (1984)
Flavour contribution of wood in alcoholic beverages.
pp. 409 - 416. In : Progress in Flavour Research.
Adda, J. ed.. Elsevier, Amsterdam.
36. Swan, J.S., Howie, D. & Burtles, S.M. (1981)
Sensory and instrumental studies of Scotch whisky flavour.
pp. 201 - 223. In : The Quality of Food and Beverages: Vol. 1
Charalambous, G. & Inglett, G. eds. Academic Press, New York.
37. Sharp, R. (1983)
Analytical techniques used in the study of whisky maturation.
Curr. Dev. Malt. Brew Dist., 143 - 157.
38. Maarse, H. (1986)
Wine characterisation and screening. pp. 18 - 44.
Proc. 10th Wine subject day on quality assurance in the wine industry. AFRC Institute of Food Research, Reading, England.
39. Maarse, H. & Visscher, C.A. (1988) (in preparation)
Volatile compounds in alcoholic beverages, qualitative and quantitative data. TNO-CIVO Food Analysis Institute, Zeist.
40. Cantagrel, R. (1986)
Application de l'analyse multidimensionelle a la caracterisation des cognacs par rapport aux autres eaux de vie du vin et alcohols de vin. Paper presented at the XIX Congr. Intern. Vigne Vin, Santiago, Chile, November, 1986.

2

Taints - causes and prevention

Derek G. Land
Taint Analysis and Quality Services,
8 High Bungay Road, Norwich, NR14 6JT, UK

WHAT IS A TAINT?'

A taint is defined in British Standard 5098(1975) as a taste or odour foreign to the product. It is a perceived attribute, or attributes, caused by the presence of chemical contaminants; these produce undesirable or different(foreign) changes in flavour (odour, taste, after-taste) which reduce the acceptability of a product. Taints are usually caused by trace (parts per million, ppm, or more often, parts per billion(10^9), ppb) levels of highly potent odourous substances; these levels must be sufficient to be perceived, or they do not produce taints. This means that they are present at above threshold concentration level for at least some people.

THRESHOLDS

The odour or taste threshold, usually defined as the concentration in a specified medium which is detected by 50% of a specified population, or, more precisely, as the concentration detected on 50% of occasions by each of a specified group of individuals, is central to the understanding of taint. Any odorous contaminant present at below its threshold for any individual will not produce a taint for that individual, although it is still a contaminant. All taints are contaminants, but not all contaminants are taints. Thresholds are both widely used and, equally, misused. The most common mis-use is to regard the threshold as an inherent property of the chemical. It is not, because the threshold for any substance will vary with the medium in which it is present, ie, the food, drink or other product (eg cigarettes, packaging, pharmaceuticals, toiletries), and between different people. This was beautifully demonstrated by the data of Zoeteman & Piet(1973) in studies of taints in drinking water in The Netherlands.

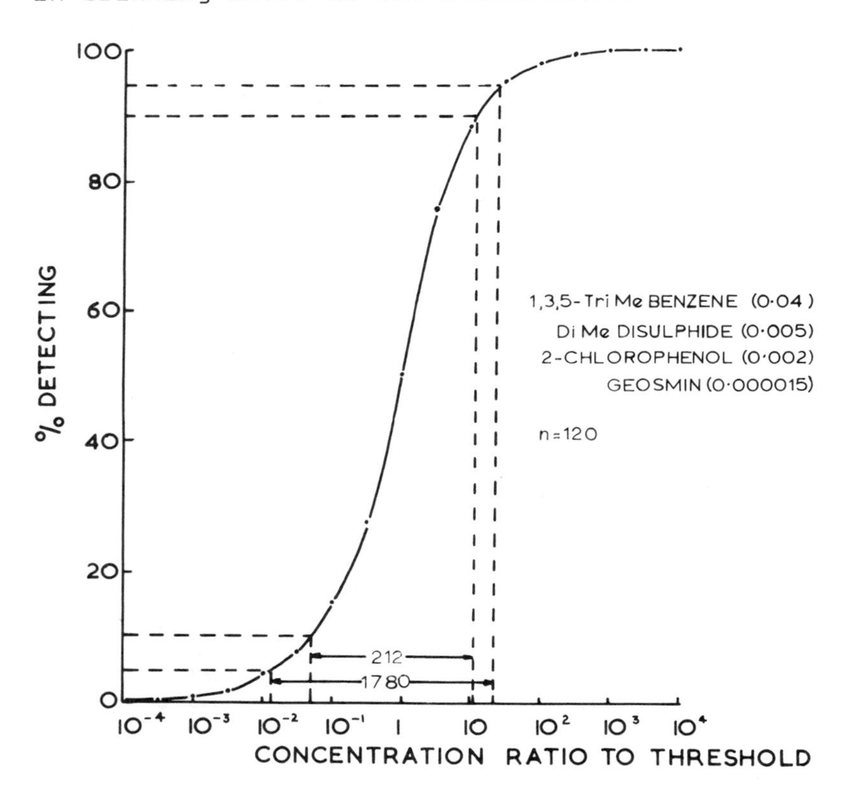

Figure 1. Cumulative distribution of sensitivities of 120 people to four different tainting substances in drinking water. Data derived from Zoeteman & Piet(1973). Concentrations are expressed as ratios of 50% population thresholds (ppm). Reproduced from Williams & Aitkin (1983) by permission of Society of Chemical Industry/Ellis Horwood Ltd.

Fig. 1 shows the normalised cumulative distribution of the concentrations of four substances differing widely in chemical structure and in mean threshold for 120 people. The centre point of each original curve, which is the 50% threshold concentration, was used as the base from which to express each test concentration as a ratio of threshold concentration. This allowed all four curves to be plotted from a common centre point, and it is clear that they form a smooth, normal cumulative distribution.

The important observation about the data is not their uniformity, but the extent to which the extremes of detectable concentration diverge. The threshold concentration difference between 10% detecting and 90% detecting is over 200-fold, whereas that between 5% and 95% is nearly 1800-fold. The most sensitive few people are about 2000 times more sensitive than the least sensitive few. This clearly demonstrates why most complaints about taint arise from only a small proportion of those buying a particular batch of tainted product - in most cases the level of contamination is such that only the more sensitive consumers detect it as taint.

This large range of sensitivity between individuals also explains many "inconsistencies" in reactions of different people to the same sample. Some will reject samples which others will happily accept. If someone complains and you cannot detect anything wrong this may be your insensitivity to the particular taint rather than their imagination; conversly if you detect a taint and others do not , you may be one of the most sensitive 5 or 10% of the population. A similar variation in response to normal flavour substances is probably also a substantial factor in determining likes and dislikes, and thus variations in choice or purchase behaviour for food and drink.

Thresholds also vary greatly between different chemical structures (Fig.1), and also between those which are very closely related. Table 1 illustrates the variation in threshold between three musty-smelling and closely related polychloroanisoles in water, and the large influence of the medium in which they are present. The 50% threshold for 2,4,6-trichloroanisole in water is 1mg, or about 0.001cc in 5 million gallons of water - a very potent tainting substance.

Table 1. Effect of structure and medium on polychloroanisole odour thresholds (ppm).

Medium	2,4,6-tri-	2,3,4,6-tetra-	Penta-
Water	3×10^{-8}	4×10^{-6}	4×10^{-3}
Tea	-	1.5×10^{-8}	-
Blancmange	-	8×10^{-10}	-
Beer	7×10^{-6}	-	-
Wine	1×10^{-5}	2.5×10^{-2}	-
20% EtOH/water	5.6×10^{-6}	1.4×10^{-5}	-
40% EtOH/water	3.6×10^{-5}	-	-
Cork	(1×10^{-2})	-	-

Sources-Curtis et al,1972; Jewel,1976; Maarse et al,1987; Tanner et al,1981; Zehnder et al,1984,() estimated.

In drinks the threshold concentration increases with ethanol content and flavour; it exceeds that in water by approximately 200-fold in beer (about 5% ethanol), 300-fold in wine (11%) ethanol), 200-fold in 20% ethanol in water, and over 1000-fold in 40% ethanol in water. This can be explained by the greater solubility in ethanol, and therefore decreased vapour pressure. There are no published figures for distilled beverages, but it would be reasonable to expect thresholds around 10^{-4}ppm in the undiluted spirit, although this would decrease sharply with dilution for drinking, or for sensory testing. The increased potency in tea and blancmange is very difficult to explain.

The real odour threshold is that in air, for this is always the transfer medium to the olfactory epithelium, but, especially for highly potent odours, it is very difficult to measure directly without major errors or artefacts. Maarse et al (1987) provide excellent advice on methodology, but some very low threshold data should be treated with caution.

TAINT CHARACTERISTICS

Thus a taint is a perceived foreign attribute resulting from the interaction of a released product contaminant with the various sensory receptors of the perceiving individual. The contaminant is usually a trace of a chemical with potent sensory properties, is usually perceived by only a minority of people, and it usually only occurs sporadically.

Often the taint is at a level at which only a small proportion of people can easily detect or recognise it; although many more will find reduced acceptabilty. This contributes to the apparent sporadic appearance, and makes detection difficult, although it may limit the impact of the problem. However, few producers would relish loss of even 10% of their market. The sporadic and often short-lived character of taints also makes investigation difficult; for the taint may have disappeared by the time its presence has been confirmed. Careful and systematic management is required, to ensure that even a short-lived taint can be traced, and this requires knowledge and understanding of how taints are formed.

TAINT FORMATION OR OCCURRENCE

Taints can occur sporadically from many sources, and have many perceived characteristics. The sources may be direct, by tainting contaminants of one or more of the many production inputs passing through the system into the product; they may be indirect, by non-tainting input contaminants which are subsequently converted to taints during production or in packaging or storage. They may also occur as a result of slight deviations from normal processes, eg by excess energy input as heat or light.

Direct sources of taint

Contact materials

The most obvious direct source of taint is by contact of inputs or product with material containing the contaminating substance. The use of polymeric cartons or bottles is a good example in products; when polystyrene became readily available as a cheap and resilient package for yoghurts, there were many problems of taint caused by diffusion of styrene monomer into the liquid contents. In production, the use of polymeric storage tanks and flexible pipes, or the use of gaskets, gland or valve seals can provide a sufficient reservoir of "foreign" substances to cause taint, usually as a result of misuse for incompatible liquids, eg in transit in tanker lorries.

Air transfer

Raw materials are also prone to contamination, either by residues of agro-chemicals, which rarely cause taint, or by aerial drift from a different operation. A relevant example of this was when grapes on the vine were contaminated by a disinfectant taint from a nearby factory coating copper wire with PVC (Tanner, 1972). The substances tainting the wine and its distilled spirit were not specifically identified other than being phenols or cresols, but almost certainly involved 6-chloro-o-cresol. This substance contaminated freshly baked biscuits up to 5 miles away, although only under conditions of atmospheric inversion (Griffiths & Land, 1973; Goldenberg & Matheson, 1975). In this case, although the major contaminant was 4-chloro-o-cresol, which has a threshold of 3ppm in semi-sweet biscuit, the taint was caused by much lower levels of 6-chloro-o-cresol, which has a threshold in biscuits of only 5×10^{-5}ppm. This substance has caused many taints in a wide variety of foods.

Water

Water is a, if not the, major input in distilled or any beverage production. If it is contaminated with a taint, eg the earthy geosmin produced by some actinomycetes, streptomycetes and algae (Gerber, 1968; Rosen et al, 1968; Medsker et al,1968), it can occasionally contaminate products; this seems to be rare, probably because water authorities are obliged to make strenuous efforts to remove foreign tastes from water (Gauntlett & Packham, 1973). Geosmin has caused taint in a malt distillate (Bemelmans & te Loo, 1976), although in this case it was thought to have been produced on rye stored at too high a moisture content.

Traces of phenols in ground water supplies have long been known to cause disinfectant taint problems on chlorination (see below).

However, any other production input which is tainted is likely to cause direct taint in the product; their sources are usually relatively easy to find after the event.

Indirect sources of taint

Indirect taints occur only as a result of two or

more factors coinciding; they are often very difficult to trace. In all cases the primary contaminant is not itself tainting. There are many different causes, only some of which can be illustrated here.

Chemical contaminants

A chemical contaminant can react with a normal constituent of either the product or an ingredient. Catty (tom-cat urine), or ribes (flowering currant blooms) taint is a good example. It can occur when trace impurities in certain solvents used in non food-use certified paints, surface coatings, adhesives, inks or wetting agents diffuse into commodities and react with traces of H_2S normally present after heating many protein-containing foods (Pearce et al, 1967; Patterson & Rhodes, 1967). The branched-chain keto-thiols or thiols formed have very low thresholds. If the raw grain was contaminated before use, this could easily happen during kilning malted barley or in the distillation, although the thiols would then be removed by the copper of the still. Beware of stainless steel or other substitutes for stills, for catty taint may well emerge.

Disinfectant taint is a very common example of reaction of a contaminant with an applied chemical. Traces of phenolic substances from external disinfection can readily react with traces of applied chlorine, either present in municipal water supplies, or as residues from hypochlorite pipe and tank cleansing, to form extremely potent disinfectant taints (Burtschell et al, 1959). For example in milk, 100-200 ppb of chlorine will produce a pronounced disinfectant taint if the milk contains only 2.5 ppb of phenol (Schlegel & Babel,1963). The taint is just detectable at 50 ppb and disappears above 250 ppb, although the concentration required to produce this "window" increases with the phenol level. This explains the sporadic nature of such taints, but also illustrates how easy it is to produce if phenolic disinfectants are ever used even in the vicinity of traces of chlorine or hypochlorite. Natural phenols, such as salicylates produced from some types of decaying vegetation, have also long been known, amongst many others, as potent sources of disinfectant taint when water supplies are chlorinated (Adams, 1931). The problem is frequently associated with under-chlorination, or contact of unchlorinated aqueous inputs, eg juices, private wells or springs with chlorinated supplies or residual hypochlorite from disinfection of tanks, pipes etc. Similar problems can occur with iodophors used for cleansing.

Methyl bromide fumigation of grain is an example in which a deliberately applied chemical reacts with a normal constituent of a commodity to produce a taint. This may be relevant to some distilled beverages which use milled grain, although I do not know of any published example of such a taint. The methyl bromide reacts with protein-bound methionine in damaged grain to form the S-methyl sulphonium derivative. On heating this produces excess dimethyl sulphide (Williams & Nelson, 1968), which is perceived as an over-cooked cabbage note in food, reducing acceptability (Land et al, 1981).

Musty or corky taints occur when a chemical contaminant is microbiologically converted, by normally indiginous organisms, to the tainting substance. The musty chloroanisoles, already referred to above, are of this type. Corks, used to close bottles of alcoholic beverages, occasionally develop a smell which gives a "corky" taint to wine (Tanner et al, 1981), or a heavy, walnut-like after-taste to brandy (Maarse et al, 1987). This taint is caused by microbial methylation of 2,4,6-trichlorophenol to form the corresponding anisole, a conversion first shown as a cause of musty taint in chickens (Dennis et al, 1975, Land et al, 1975) and subsequently shown to be caused in corks by a combination of chlorine bleaching with microbial methylation of the resulting chlorophenol (Zehnder et al, 1984). Irradiation sterilization of the bleached corks prevented anisole formation. Chloroanisoles have caused widespread taints in many situations (Maarse et al,1985; Whitfield, 1986), mainly because of the extremely widespread use of polychlorophenols as fungicides in wood and paper products. The treated packaging for empty gin bottles even caused a musty taint in the subsequently filled bottles (Maarse et al,1985).

Oxidation taints are caused by exposure of products to oxygen, light or both. Examples are the fatty, rancid painty notes in spirit derived from grape marc (Williams & Strauss, 1978), and "sunlight" flavour in champagne (Maujean & Seguin, 1983) or even in beer in brown bottles (Kavanagh et al, 1983). The reactions and products are different in each of these cases.

Microbial contaminants

The effects of contamination of wort by wild yeasts are too well-known in the fermentation industries to require more than passing mention (eg, Bertrand, 1983); they also produce effects other than taints produced by their normal metabolic products. Contamination of wet or poorly stored grain also produces off-odours which will contaminate products (Stawicki et al,1973).

Other, less well-known effects of microbial contaminants, are the formation of a "geranium" flavour from added sorbate in wine by lactobacilli (Crowell & Guymon, 1975), and of "Charantais rancio", a rancid flavour in vintage cognacs, thought to be caused by oxidase enzymes liberated from moulds (Lafon, 1976).
This account of sources of taint is by no means complete, partly because it would require a treatise to adequately cover what is published, and partly because much information remains confidential and cannot be backed by published reference; however experienced investigators do use such information in their cases.

PREVENTION

Whatever the origin, a taint causes minor or major disruption, and incurs cost. If a tainted product is distributed to consumers, it can be disastrous in terms of both short-term rescue costs and long-term market share. Many cases in which losses have exceeded £1M are known, but are not publicised. It is also rare for taints to be satisfactorily removed, even at substantial cost. The only logical strategy is therefore prevention, risk minimisation, and a clear, agreed, and enforced system of specification and monitoring to detect any taint at the earliest possible stage. Although in some cases risks may be reduced by general limits on allowable levels of, eg, chlorophenols in contact material, as suggested by Maarse et al,(1987), taints arise by such variable routes and sources that responsibility for action must be at a more local level, between suppliers of inputs and producers, and between producers and their customers.

Malt whiskey production - inputs and variables

As a relevant example to outline the approach, malt whiskey production can be used, although not because it has more than its share of taints than other products. The first step is to identify inputs at all points in the process (Table 2).

Table 2. Input factors in malt whiskey production

Normal inputs	Plus ?	Secondary inputs	Marginal effects
Barley	Microbes	Cleansers	Drought
	Pesticides		
	Storage odours	Maintainance	Climate
Water	Chlorine	Repairs	
	Additives		
Air	Effluents		
Smoke			
SO_2			
Yeast	Wild, Mutants	INTERACTIONS	
Barrels			
Bottles			

Input specifications

Each normal input has the potential to carry a direct taint. All except the smoke can be specified as being free from unusual odour, and even the peat used to produce the smoke could be specified as not producing smoke of unusual smell. Some of these inputs may already have such a specification, eg, barley, barrels or casks, and bottles, but others will not. It is essential to know if or when sources of any inputs are changed in <u>any</u> way.

Each input also has the potential to carry contaminants which can produce indirect taint; some are listed under "Plus ?", but the main hazard is from interactions. Again, some potential factors will already be covered by specification, but others will not. For example, if the water is normally taken from a spring or upland stream, is it always checked for occasional adventitious contamination such as emergency fuel discharge from aircraft? Only traces of kerosine are needed to taint surface water from the area, although it might not emerge from a spring until some time after the event. If supplementary water is taken from domestic supplies at times of drought, it will certainly be chlorinated, and will probably contain other additives. Are procedures specified to check for sensory consequences? If maintainance or building occurs nearby, does it have adverse effects from tar boilers, diesel compressors, or spillages on the air drawn over the germinating barley? If the barley silo or the distillery is in a valley, are there distant sources of air pollution which could travel for many miles, with little dilution, under conditions of atmospheric inversion?

There are many more questions such as these, which can, and should, be answered by positive action rather than by default after problems have occurred. It is not possible to generalise to all situations, for even within a highly specialised and localised product such as malt whiskey, there is great diversity of situations. The extended situation for production of other distilled spirits shows even greater diversity of input conditions. However the principles are clear.

Variables, criteria and control

After identifying the inputs, the variables which are known to influence them, and the processes should be defined and specified. For example, the extent to which germinating barley rootlets are allowed to grow, the microbial contaminants associated with the malt, and residual SO_2 levels are all known sources of taint (Watson, 1983). All biological materials vary with many uncontrollable factors, but the processes involved in processing can at least be followed, and existing knowledge used to minimise undesirable variations. This is done to a large extent in the fermentation, although further control to reduce the bacterial formation of substances such as hydroxypropanal or butyric acid (Watson, 1983) is probably possible.

The next step is to establish criteria for specifications, whether for external or internal supplies, and for production variables. To be effective, any specification must be capable of testing, to ensure that it is within limits, and, ideally, of measurement to indicate trends which will forewarn of possible trouble in future. This requires agreement between parties on both the specification and the method of test. In some cases this can be expressed as chemical limits, eg the levels of polychlorophenols in wood or packaging for long distance containerised transit (Maarse et al, 1987), but in general sensory specifications are adequate if backed up with careful production specifications, good housekeeping and tight control.

The last requirement is to ensure that a sensible and effective level of testing is maintained. This requires good management, well- motivated and -informed staff, and an easily maintained system of testing and records.

CONCLUSIONS

Taints are caused by trace contaminants which are potent sources of foreign smell or taste. They usually occur rarely, are sporadic when they do occur, and are only detected by a minority of the population. However when they do occur they are usually highly disruptive, and costly in short term rescue, loss of product, and in future market share.

Many causes are now well understood, although often costly to investigate. However it is now possible to design and implement specifications and prevention systems which reduce risks, and improve both the speed of detection and containment of costs of any taints which do occur.

REFERENCES

Adams,B.A. Substances producing taste in chlorinated water-Part1. Water and water engineering,1931,33,109-113.

Bemelmans,J.M.H. & te Loo,N.L.A. Analysis of an earthy taint in a malt distillate. In: Proc. 4th Nordic Symposium on sensory properties of foods,Skovde, Sweden, March 18-20,1976.

Bertrand,A. Volatiles from grape must fermentation. In: Flavour of distilled beverages (ed Piggott,J.R.) 1983,pp93-109.

British Standard 5098, Glossary of terms relating to sensory analysis of food, British Standards Institute, London, 1975.

Burtschell,R.H., Rosen,A.A., Middleton,F.M. & Ettinger,M.B. Chlorine derivatives of phenol causing taste and odor. J. Amer. W.W. Assoc.,1959,51,205-214.

Crowell,E.A. & Guymon,J.F. Wine constituents arising from sorbic acid addition and identification of 2-ethoxyhexa-3,5-diene as a source of geranium-like off-odour. Amer.J.Enol.Viticult.,1975, 26,97-102.

Curtis,R.F., Land,D.G., Griffiths,N.M., Gee,M.G., Robinson,D.,Peel,J.L., Dennis,C. & Gee,J.M. 2,3,4,6-tetrachloroanisole association with musty taint in chickens and microbiological formation. Nature,1972,235, 223-224.

Dennis,C, Mountford,J., Land,D.G. & Robinson,D. Changes in the microbial flora, chlorophenols and chloroanisoles in broiler house litter during a chicken rearing cycle. J.Sci.Fd.Agric., 1975,26,861-867.

Gauntlett,R.B. & Packham,R.F. The removal of organic compounds in the production of potable water. Chemy. Ind.,1973,812-817.

Gerber,N. Geosmin, from micro-organisms, is trans-1,10-dimethyl -trans-9-decalol. Tetrahedron Lett.,1968,25,2971-2974.

Goldenberg,N. & Matheson,H.R. Off-flavour in foods, a summary of experience:1948-74. Chemy. Ind.,1975,551-557.

Griffiths,N.M. & Land,D.G. 6-Chloro-o-cresol taint in biscuits. Chemy. Ind.,1973,904.

Jewel,G.G., The relationship between disinfectant and musty taints in foods and the presence of chlorophenol derivatives. BFMIRA Tech. Circ.,1976,No. 616.

Kavanagh,T.E., Skinner,R.N. & Clarke,B.J. Light-struck flavor formation in beer in illuminated display cabinets. Brewers Digest,1983,58(2),44-46.

Lafon,J. Origin and formation of Charentais Rancio off-flavour in vintage cognacs. Ann. des Falsif. et de l'Expert. Chim.,1976, 69(739),315-318.

Land,D.G., Gee,M.G., Gee,J.G. & Spinks,C.A. 2,4,6-Trichloroanisole in broiler house litter: a further cause of musty taint in chickens. J.Sci.Fd.Agric.,1975,26,1585-1591.

Land,D.G., Griffiths,N.M., Reynolds,J. & Davies,A.M.C. Dimethyl sulphide- an undesirable food component in chicken? In: Criteria of food acceptance (eds Solms,J.& Hall,R.L.), Forster Verlag, Zurich,1981,pp355-358.

Maarse,H. & van der Berg,F. Current issues in flavour research.
This volume.

Maarse,H., Nijssen,L.M. & Jetten,J. Chloroanisoles, a continuing story. In: Topics in flavour research(eds Berger,R.G., Nitz,S. & Schreier,P.)Eichhorn, Hangenham,1985,241-250.

Maarse,H., Nijssen,L.M. & Angelino,S.A.G.F. Halogenated phenols and chloroanisoles: occurrence, formation and prevention. In: Proc. 2nd Wartburg Aroma Symposium, Nov,1987, (ed Rothe,M.)Akademie Verlag, in press.

Maujean,A. & Seguin,N., Sunlight flavours in champagne wines,3;Photochemical reactions responsible. Sciences des Aliments,1983,3(4),589-601.

Medsker,L.L., Jenkins,D. & Thomas,J.F. Odorous compounds in natural waters. An earthy-smelling compound associated with blue-green algae and actinomycetes. Environm. Sci. Technol. 1968,2,461-464.

Patterson,R.L.S. & Rhodes,D.N. Catty odours in food: their production in meat stores from mesityl oxide in paint solvents. Chemy. Ind.,1967,2003-2004.

Pearce,T.J.P., Peacock,J.M., Aylward,F. & Haisman,D.R. Catty odours in food: reactions between hydrogen sulphide and unsaturated ketones. Chemy. Ind.,1967,1562-1564.

Rosen,A.A., Safferman,R.S., Mashni,C.I. & Romano,A.H. Identity of odorous substance produced by Streptomyces griseoluteus. Appl. Microbiol.,1968,16,178-179.

Schlegel,J.A. & Babel,F.J. Flavors imparted to dairy products by phenol derivatives. J.Dairy Sci.,1963,46,190-194

Stawicki,S., Kaminski,E., Niewiarowicz,A., Trojan,M. & Wasowicz,E. The effect of microflora on the formation of odors in grain during storage. Ann.de Technol.Agric.1973,22,449-476.

Tanner,H. Ein fall von weinkontamination durch industriebedingte emissionsproducte. Mitt. Lebensm. u. Hygeine,1972,63,60-71.

Tanner,H., Zanier,C. & Buser,H. 2,4,6-Trichloroanisole: eine dominierende komponente des korkgeschmackes. Schweiz. Zeitschr. f. Obst- und Weinbau,1981,117,97-103.

Watson,D.C.,Factors influencing the congener composition of malt whiskey new spirit. In:Flavour of distilled beverages (ed Piggott,J.R.), Ellis Horwood,1983,79-92.

Whitfield,F.B. Food off-flavours:cause and effect. In:Developments in food flavours (eds Birch, G.G. & Lindley, M.G.) Elsevier Applied Science, London,1986,249-273.

Williams,A.A. & Aitkin,R.K.(eds). Sensory Quality in Foods and Beverages, Ellis Horwood,1983,p.23.

Williams,M.P. & Nelson,P.E. Kinetics of the thermal degradation of methylmethionine sulfonium ions in citrate buffers and in sweetcorn and tomato serum. J. Food Sci.,1974,39,457-460.

Williams,P.J. & Strauss,C.R. Spirit recovered from heap-fermented grape marc:nature, origin and removal of the off-odour. J. Sci. Fd. Agric.,1978,29,527-533.

Zehnder,H.J., Buser,H.R. & Tanner,H. Zur entstehung des korktons in wein und dessen verhinderung durch die behandlung der flaschenkorken mit ionisierender strahlung. Deutsche Lebensm. -Rundschau,1984,80,204-207.

Zoeteman,B.C.J. & Piet,G.J. Drinkwater is nog geen water drinken. H_2O,1973,6,174-189.

3

Sensory analysis of alcoholic beverages

Peter Dürr
Swiss Federal Research Station for Fruit-Growing, Viticulture and Horticulture, CH-8820 Waedenswil, Switzerland

ABSTRACT

Sensory analysis requires an instrument, scales, methods, a technique and a skilled analyst. It starts with a clear analytical question. What should be measured? This can be a product variable or a consumer response. The quality of a product has two sources of variables, the product and its taster. Experts and consumers have individual, hardly changing and tough concepts about alcoholic beverages. Ethanol enhances the volatility and the retronasal effect of odour compounds. The importance of ethanol concentration in samples must be stressed. Diluted samples are less fatiguing for the tasters but different in characteristics. A few aspects of current sensory research are presented to document the state of the art.

1. INTRODUCTION

Modern sensory analysis, based on psychology, psychophysics, physiology and food science, is used to measure:

- the occurrence, size and type of differences between two food samples
- the increase or decrease of product parameters by ranking
- several product characteristics by descriptive

analysis
- odour and taste intensities by magnitude estimation
- the occurrence, size and type of products taints
- the hedonic response to different stimuli
- consumer attitude towards familiar or new products and ideas.

Sensory analysis requires an instrument, which is always a group of people, scales, clearly defined methods, application by a skilled analyst and a certain amount of space, time and money. Sensory analysis, like any other analysis, should start with a clear formulation of the analytical question. However, more often than not the approach to sensory work is undefined. What is to be measured? It is usually a quality variable. The quality of a product has two sources of variables, the product and its taster. The product has variables such as composition, colour, flavour, pH and its taster has variables such as expectations, experience, linguistic talent and current disposition.

2. THE SENSORY INSTRUMENTS

The human instruments can be divided into three types: the consumer group, the research panel and the small expert group.

Consumer groups are often used to measure product characteristics, but are also themselves objects of investigation, for example when measuring consumer preference or attitude toward a product. Since consumer groups often show a wide variation in response, relatively large groups are used. The final number can be determined in a stepwise procedure. It is difficult to control the performance of the individual. The working place of a consumer group should be close to consumer reality.

Due to statistical requirements a research panel consists of 15-25 assessors, highly trained on methods and technique. Familiarity with particular products is unnecessary, but is acquired through experience. Such a group shows only a small response variation. The individual performance and the work in the sensory research laboratory are both easy to control. However, a research panel is too small to measure consumer response.

The expert is characterised by excellent product knowledge and training in sensory techniques. The response variation is small to medium, therefore the group must consist of at least 3 experts with a strict control and feedback of individual performance. Experts have an extended vocabulary to describe product characteristics but also severe concept problems, such as the use of different words for one stimulus or the same word for distinctly different stimuli.

3. THE SCALES

Any measurement needs a scale. Sensory analysis uses scales based on human characteristics. Humans respond more exponentially than linearly to sensory stimuli. Therefore, three types of scales with their own set of statistical methods are appropriate:

The nominal scale is related to the "equal or different" question. Is there a perceptual difference between two samples? The answer is yes or no; it is very precise but not very informative.

The ordinal scale is related to the "more or less" question. It is more informative but doesn't provide quantitative data. It is not linear and therefore connected to non-parametric statistics. The ordinal scale is the most popular scale in sensory analysis and used for example in ranking samples along a given product attribute or along a hedonic dimension (preference). The various point scales are often misused as the equal intervals on these scales are disregarded [1], which gives data on the ordinal level.

The ratio scale is related to the "how many times more or less" question. It provides quantitative data but is less precise. An example is the sensory measurement of odour intensity of beverages [2]. The ratio scale is commonly used in instrumental or chemical measurements.

4. THE METHODS

The sensory methods are described in almost any text book on sensory analysis. PIGGOTT [3] compiled a concise review of modern sensory analysis. Some of the methods were developed pragmatically, others derive from psychophysics. O'MAHONEY [4] presented the psychologist's approach to sensory statistics, recommended reading for a sensory analyst. A detailed laboratory manual of sensory analysis has been written by JELLINEK [5].

The methods can be grouped into difference tests, ranking procedures, verbal descriptions, profiling methods and intensity measurements.

Difference tests

The difference tests are known as paired comparison or duo, triangle, duo-trio tests etc. They were pragmatically developed and are somewhat ambiguous but are now under revision by the psychophysicists

[6,7]. The sensitivity of consumer groups or research panels is variable and difficult to control. The error probability of the panel result depends on the size of the panel. There are further questions on how a significant difference is perceived: visually, nasally and/or orally and how it is characterised. To measure the size of a difference, another dimension is required. The assessors are asked how sure they are about their answers. This degree of confidence is related to the size of the difference and calculated as the r-index [8]. Another practical problem is often the relation between the difference found and the product history.

Ranking and scoring procedures

In ranking, 2-5 samples are presented together to be ordered according to a given criterion. Criteria are colour, odour or taste characteristics, or preference. In scoring, more samples can be evaluated if they are presented one by one in random order. Only a few criteria should be scored with short point scales. Nonparametric KAHAN/KRAMER tables [9] are used to determine the significance of such data.

Descriptive analysis, profiling

Verbal analysis of food and beverage samples is the most time consuming and demanding part of sensory analysis, but it can provide highly informative and complex data. Product characteristics and their descriptive terms have to be defined. This is best done in co-operation with the panel to be trained. Training is word concept formation. Problems arise especially when product experts are testing products of high hedonic value like distillates. It has been shown that experts can even fail to understand their own descriptions [10]. To overcome the problems with such strongly fixed personal vocabularies, WILLIAMS [11] proposed free-choice profiling. The assessor works with his personal vocabulary and a short point scale to measure the intensity of attributes of his choice. This information is used to give each sample a position in a multidimensional space. By using rotational, stretching and shrinking techniques, it is possible to match each assessor's spaces to others and yet maintain the relationship between samples. The difficulty is to relate the axes of such multidimensional spaces to verbal attributes, which find general acceptance.

Multivariate statistics are used to find correlations between the variables assessed with profiling and other variables measured chemically or physically on the same samples.

Intensity measurement

The perceptual intensity of colour, odour and taste can only be measured with sensory means. The function between the concentration of a stimulus and the perceived intensity is basically described by Stevens' law [12],

$$I = S^n$$

where I = perceived intensity,
S = stimulus concentration, and
n= exponent of power function.

Stevens' law is valid between the sensory limits of threshold and saturation. The ideal method of measuring intensities is magnitude estimation [12]. The best instrument is a group of about 25 persons, clearly instructed as to their task. This instrument is calibrated with samples of known stimulus concentration like any other instrument. The result is expressed as the median or the geometric mean of all answers.

5. THE SENSORY TECHNIQUE

Sensory analysis requires skilled and experienced staff to select, motivate and control assessors and to run the tests properly. Place of testing and time of day are critical factors. It is important to control the amount of information available to the assessors and the communication between staff and assessors. Assessors have to work independently with properly coded samples. Whenever possible, the samples are presented one by one in random order to avoid neighbour effects. If necessary, double blind procedures are applied. Under- and over-load of assessors should be avoided. The following example shows the effect of overload.

6. EXAMPLE: EXPERIMENTS WITH PEAR DISTILLATES

A service laboratory organised an experiment with three different pear juice treatments prior to fermentation and distillation. Four distilleries were involved in this industrial scale experiment. The distillates were analysed in three different laboratories. Part of the analytical work was a sensory evaluation with product experts.

The original test sheet contained this information for the assessors:

Experiments in distillate production
Ranking method
The samples are diluted to 30 %Vol. ethanol
You get 4 series with 3 samples each
rank according to: (rank 1 is the best sample)
1. odour type
2. odour intensity
3. taste type
4. taste intensity
5. general impression
comment on your decisions

From a panel with 22 product experts, three refused to work with this sheet and eight experts provided incomplete data. The panel was strongly overtaxed by unclear information and by too many questions. Then the test sheet was revised by a sensory analyst and the number of samples reduced by two since these had clear taints.

Evaluation of pear distillates 30%Vol.
s = 4 x 2 or 3
rank according to
1. odour preference
2. taste preference
comment on your decisions. It is important!

With this sheet there were no complaints from the panel and data were complete, with sufficient descriptions for data interpretation.

7. CURRENT SENSORY RESEARCH

A few topics of current sensory research applicable to practical analytical work can serve to illustrate the state of the art. Odour is perceived as such by the nose with the olfactory mucosa, but taste on the tongue is simultaneously perceived with volatiles in the nose. So there are nasal, retronasal and oral effects to separate [13]. The time-intensity relation [14] is another aspect of perception, important with alcoholic beverages or modern sweeteners. The concept of acceptance incorporates both preference and attitude. Preference is regarded as the degree of liking or disliking. Attitude is closely linked to beliefs. TUORILA [15] gives a practical example concerning sweetness in product development. Language is a common information carrier with some drawbacks. LEHRER [16] found a lack of communication among wine tasters. Further experiments with wine experts came to equally disillusional results [10,17]. How far training can help to improve the tasters performance is an open question. The study of concept formation mechanisms may provide an answer.

8. SPECIFIC PROBLEMS OF HIGHLY ALCOHOLIC BEVERAGES

Alcoholic beverages are not consumed as a nutritive food, but as a drug for mainly pleasure or relaxation. Therefore such products have a very pronounced hedonic impact. Experts and consumers describe what they like and dislike and not the product itself. They have individual, stable and tough concepts about product parameters.

Ethanol has an irritating and strong retronasal effect. Decreasing the ethanol content decreases the solubility, volatility and the retronasal effect of odour compounds, which has to be taken into account in the development of dealcoholised beverages. With distillates, the retronasal perception of volatiles after swallowing or spitting the sample is an important quality aspect.

The ethanol concentration of samples is an important factor in practical testing. As a rule, distillates are tested at the product concentration normal for consumption. Diluted samples are less fatiguing for the tasters. But diluting changes the product character, however it may be helpful to find irregularities and taints.

REFERENCES

1 Cloninger, M.R., Baldwin, R.E. and Krause, G.F., Analysis of sensory rating scales, J. Food Sci., **44**(5), 1225-1228 (1981)
2 Duerr, P., Sensorische Beurteilung der Intensitt von Fruchtsaftaromen, Ber. Intern. Fruchtsaftunion, 15, 155-161 (1978)
3 Piggott, J.R., Editor, Sensory Analysis of Foods, ISBN 0-85334-272-5, Elsevier Applied Science Publ., London (1984)
4 O'Mahoney, M., Sensory Evaluation of Foods: Statistical Methods and Procedures, ISBN 0-8247-7337-3, Marcel Dekker, New York (1986)
5 Jellinek, G., Sensory Analysis of Food, ISBN 0-89573-401-X, Ellis Horwood, Chichester, England (1985)
6 Frijters, J.E.R., Blauw, Y.H. and Vermaat, S.H., Incidental training in the triangular method, Chem. Sens., 7(1), 63-69 (1982)
7 O'Mahoney, M., Wong, S.Y. and Odbert, N., Sensory difference tests: some rethinking concerning the general rule that more sensitive tests use fewer stimuli, Lebensm. Wiss. Technol., **19**, 93-94 (1986)

8 O'Mahoney, M., Short-cut signal detection measure for sensory analysis, J. Food Sci., **44**(1), 302-303 (1979)
9 Kahan, G., Cooper, D. Papavasiliou, A. and Kramer, A., Expanded tables for determing significance of differences for ranked data, Food Technol., **27**(5), 61-69 (1973)
10 Duerr, P., Wine description by experts and consumers, Proceedings second international cool climate viticulture and oenology symposium, Auckland, New Zealand, January 1988
11 Williams, A.A. and Langron, S.P., A new approach to sensory profile analysis, in Piggott, J.R., Ed., Flavour of distilled beverages, p. 219-224, Ellis Horwood Ltd., Chichester, England (1983)
12 Moskowitz, H.R., Scharf, B. and Stevens J.C., Eds., Sensation and measurement: papers in honour of S.S. Stevens, D.Reidel Publ., Dordrecht, Holland (1974)
13 Burdach, K.J., Kroeze, J.H.A. and Koester, E.P., Nasal, retronasal and gustatory perception: an experimental comparison, Perception Psyhophysics, **36**(3), 205-208 (1984)
14 Lee, W.E. and Pangborn R.M., Time-Intensity: the temporal aspects of sensory perception, Food Technol., **40**(11), 71-78, 82 (1986)
15 Tuorila, H., Hedonic responses and attitudes in the acceptance of sweetness, saltiness and fattiness of foods, Proceedings IUFoST-Symposium Zurich 1987, Academic Press, London
16 Lehrer, A., Is semantics perception-driven or network-driven? Austral. J. Linguistics, 5, 197-207 (1985)
17 Lawless, H.T., Flavor description of white wine by expert and nonexpert wine consumers, J. Food Sci., **49**(1), 120-123 (1984)

4

Consumer free-choice profiling of whisky

C. Guy, J. R. Piggott and S. Marie
Food Science Division, Department of Bioscience and Biotechnology,
University of Strathclyde, 131 Albion Street, Glasgow, G1 1SD, Scotland

ABSTRACT

Free-choice profiling has recently been used for collecting profile information on a variety of foods and beverages from laboratory panels of selected assessors, but the use of this method by a large panel of consumers has not been reported. Free-choice profiling and subsequent Generalised Procrustes Analysis were therefore used for descriptive analysis of whisky by a consumer panel, to identify the dimensions used by consumers to discriminate between whiskies. The same whiskies were profiled by a trained laboratory panel using an agreed vocabulary. The consumer data showed that the panel could discriminate between the samples, and the sample configuration could be interpreted in terms of the colour and of descriptors used by the trained panel. Free-choice profiling was found to be potentially an important and useful method for consumer testing, which could produce data of value to product development and marketing.

INTRODUCTION

For successful marketing of food products, knowledge of the consumers' perceptions and preferences is essential (Wierenga, 1986). There are many methods of testing preference and acceptance, the form used depending on the objective of the study, the product, and the consumers participating in the test (Colwill, 1987).

Conventionally sensory testing is divided into affective and analytical testing. Affective testing assesses consumer preferences for the product, whereas analytical testing encompasses descriptive sensory analysis (Reynolds, 1987). An insight into the attributes which consumers associate with acceptance can be gained by relating affective and analytical data, and hence desirable attributes can be emphasised and undesirable attributes reduced to achieve an optimum quality product (Schutz, 1987).

There are two widely used forms of conventional descriptive analysis. These are the Flavour Profile Method (Cairncross and Sjostrom, 1950) and Quantitative Descriptive Analysis (Stone et al., 1974) but both have disadvantages especially for use with untrained consumers (Williams, 1983).

To overcome some of these problems, Free Choice Profiling (FCP) was developed (Williams and Langron, 1984). FCP assumes that subjects do not differ in the number and kind of sensory characteristics they perceive in a product, but that they do differ in the way they label them (Arnold and Williams, 1986). The assessors are therefore allowed to develop their own individual vocabularies containing as many terms as they wish, and use them to score a set of samples. The terms used must have meaning for the individual, but there is no necessity to convey the meaning to any other assessor (Williams and Arnold, 1985). In this way each person generates an individual profile for each product in his own vocabulary. The individual configurations are then subjected to Generalised Procrustes Analysis (GPA) which was originally developed by Gower (1975) and has been described by Williams et al. (1981) and Langron (1983). GPA produces information on the inter-relationships between the samples and between the assessors (Williams and Langron, 1983). There is no strict selection of assessors for FCP and the training is not exhaustive. In principal, FCP should be applicable to any type of assessor whether experienced or inexperienced (Williams and Langron, 1984).

A valuable method of consumer sensory evaluation might therefore be found in the application of FCP to consumer studies. FCP should be applicable to consumers because individual word lists are developed and extensive training is not required. To give satisfactory results with FCP the consumers must simply be objective; be capable of using scales; and use the developed vocabulary consistently.

The aim of this experiment was two-fold: firstly, to determine whether FCP could be used by consumers in a consistent and meaningful way; secondly, to investigate consumer perception of smoothness and maturity in whisky. The conventionally trained whisky panel at Strathclyde University (Piggott and Canaway, 1981; Piggott et al., 1985) provided a trained panel analysis of the samples with which to compare the consumer results.

EXPERIMENTAL

The whisky brands used were: Claymore, Langs, Bells, 100 Pipers, Bells 12 year old Connoisseur, Chivas Regal and Johnnie Walker Black Label. An eighth sample was prepared by adding 0.05% caramel to Bells in order to determine how the appearance of the whisky affected perception. These brands were chosen because they are well-established, popular and sell in large quantities, and cover a range of sensory properties. These whiskies were bottled in 50 ml glass miniatures and were labelled A to H and 1 to 8 (for the 2 testing stages). The assessment of the whiskies by 100 consumers took place in the two separate stages shown in Table 1.

Table 1. Stages of the tasting sessions

	1. Group session
(i)	Instructions
(ii)	Generation of descriptive terms
(iii)	Assessment of samples
	2. At-home testing
(i)	Instructions
(ii)	Assessment of samples
(iii)	Return by post

During the group sessions, the 8 whiskies were given to each assessor to examine and list descriptive terms. Assessors were instructed to rinse their mouths with water between samples to reduce carry-over effects. It was permitted to add water to the whisky but it was stressed that the same volume should be added to every sample to ensure uniformity. The assessors were instructed to describe in their own words the appearance, aroma, and flavour of the samples. To investigate the consumers' perceptions of smoothness and maturity, these terms were added to each assessor's list. Four samples selected from the 8 whiskies were then assessed by the consumers using line scales (Land and Shepherd, 1983) under test conditions, and thus the method for scoring of the whiskies was established and practised.

For the at-home sessions, assessment of the eight whiskies was performed using the same descriptive terms and the same type of score sheets. Assessors were instructed to follow exactly the same procedure as they had during the group session.

RESULTS AND DISCUSSION

Of the 100 consumers, 97 completed the sample assessment and returned their results and, of these, 93 sets of results were usable. The number and variety of words generated varied vastly with individual assessors. An example of some assessors' descriptive terms is shown in Table 2. The intensities marked on the scales were measured and these results were then subjected to GPA.

Table 2. Examples of terms generated

Assessor 1	Assessor 2	Assessor 3	Assessor 4
smooth	smooth	dark	clean
scented	rough	smooth	sharp
bitter	mature	golden	bitter
coffee	malty	sharp	woody
dusky	golden	tart	golden
creamy	dry	peaty	deep
fishy	choking	oily	round
syrupy	sharp	burning	smooth
bland	weak	fragrant	fusty
woody	refreshing	woody	shallow
musky		musty	tobacco
burnt		sweet	rusty
thin		stale	volatile
mature		thick	mature
		greasy	raw
		diesel	caramel
		rough	nippy
		smokey	
		yellowish	
		arid	
		roast	
		earthy	
		old	
		mature	

Assessor plot

The first 2 axes of the assessor plot after GPA are shown in Figure 1. The points represent the relative distances of the assessors' individual configurations from the consensus configuration. This illustrates that the group was homogeneous with no obvious clusters of assessors or outliers. The relatively high degree of agreement shows that the consumers were assessing the samples in similar ways and perceiving broadly the same characteristics in the whiskies.

Centroid configuration

The first 3 principal axes of the centroid configuration obtained by GPA are shown in Figures 2 and 3. These 3 axes described 58% of the variance.
Axis 1 was apparently related to the colour of the samples, so the absorbance at 430 nm was measured and plotted against the Axis 1 coordinate of each whisky (Figure 4). A correlation coefficient of 0.994 was obtained showing that colour was correlated with Axis 1.

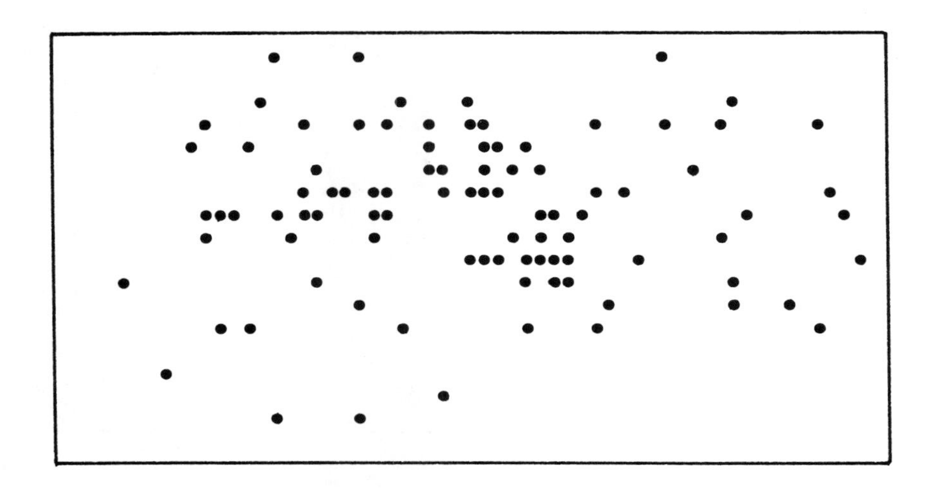

Figure 1. Assessor plot

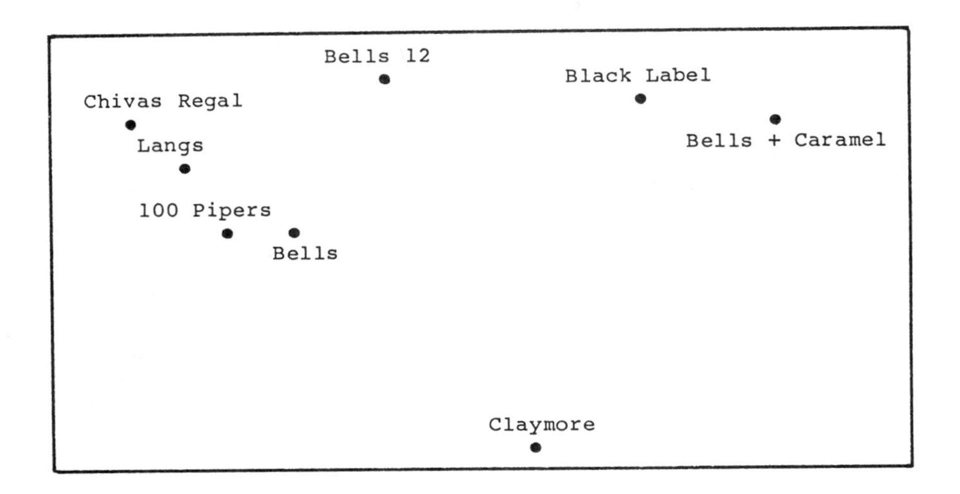

Figure 2. First (horizontal) and second (vertical) axes of consensus configuration from consumer data

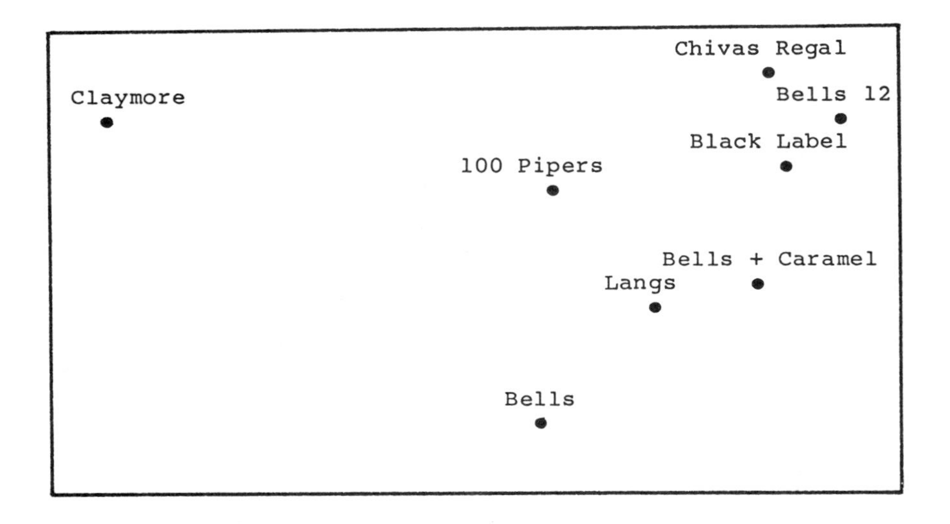

Figure 3. Second (horizontal) and third (vertical) axes of consensus configuration from consumer data

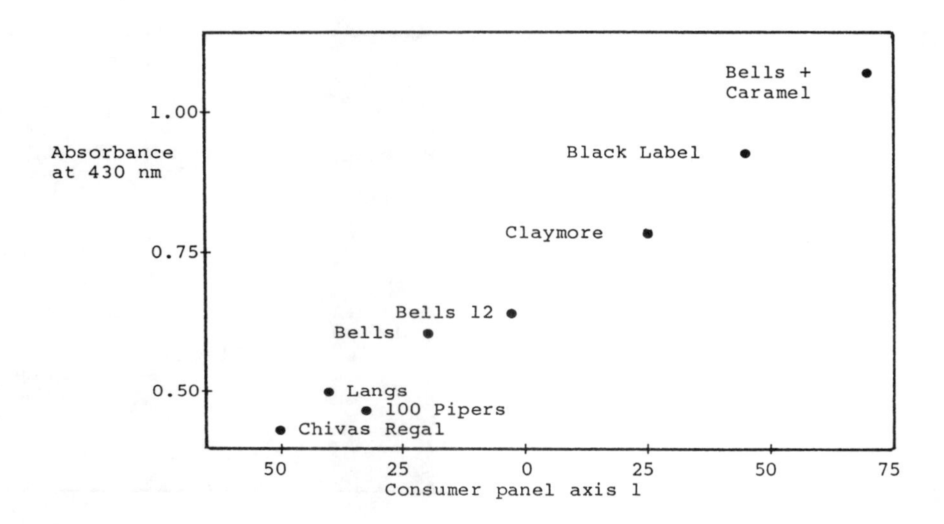

Figure 4. Sample scores on Axis 1 of the consensus configuration plotted against absorbance

It can be seen that, on Axis 2, the Deluxe Blends and the Bells plus Caramel are grouped together with high values. The Standard Blends are positioned further down the axis. These groupings suggested that Axis 2 describes the malt content or maturity of the whiskies as it clearly separates out the different types of blend. Axis 2 therefore represents the characteristics consumers associated with maturity. It is interesting to observe that the Bells plus Caramel has been placed with the Deluxe Blends on Axis 2. This suggests that the depth and tone of the colour is important to consumers when assessing maturity. By deepening the colour of the Bells a greater apparent maturity seems to have been obtained.

The meaning of Axis 3 (Figure 3) was not clear from inspection of the samples, though it apparently represented another contrast between the deluxe blends and the other samples.

Trained panel results

The trained panel data were analysed by principal components analysis without GPA; the sample configuration is shown in Figures 5 and 6.

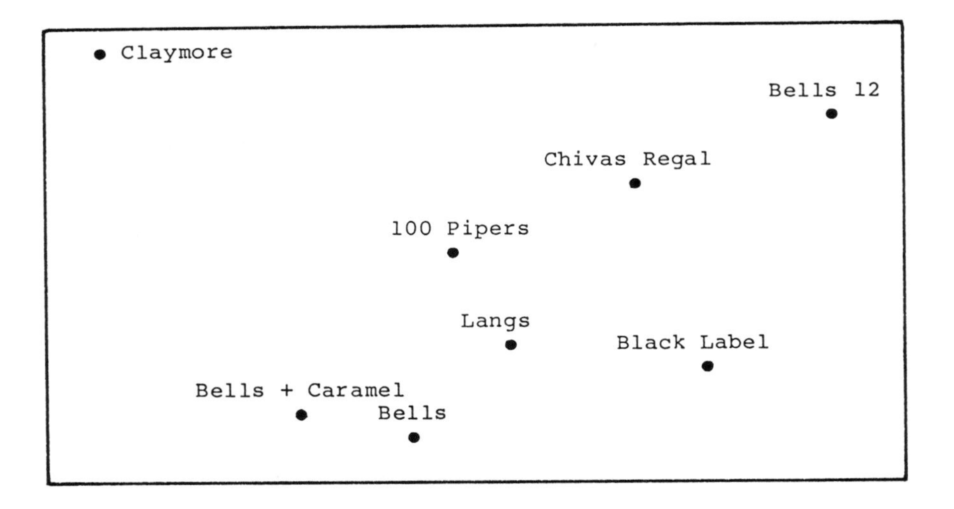

Figure 5. Sample scores on first (horizontal) and second (vertical) components from trained panel

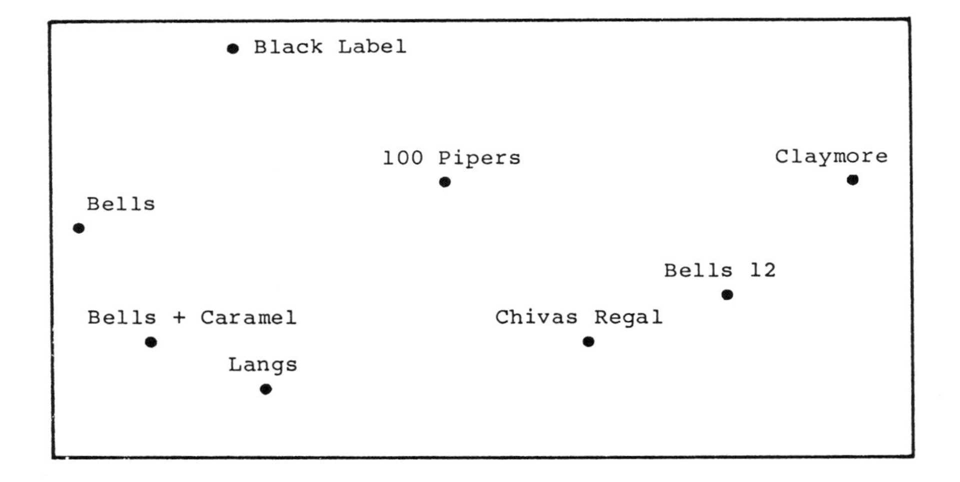

Figure 6. Sample scores on second (horizontal) and third (vertical) components from trained panel

The first two axes of the panels' sample plots were compared. Both panels have obtained similar sample groupings; this can be seen by matching Axis 2 of the consumer plot with Axis 1 of the trained panel plot. Chivas Regal, Langs, 100 Pipers, Bells, Bells 12 year old, Claymore and Black Label have all been placed in approximately the same positions by both panels. The panels vary in their positioning of Bells plus caramel, with the consumers giving it a high value on the Axis representing maturity. This difference probably arose because the trained panel does not assess colour and any differences in appearance are eliminated by testing under red lights. When the colour was not assessed, Bells and Bells plus caramel were placed together. The two panels diverged considerably when the third axes were compared; they were apparently considering different characteristics of the samples.

Consumer term usage

To help interpret the consumer dimensions, the terms which correlated with the first three principal axes were listed. The large number of assessors and the wide range of terms generated made interpretation of word usage a difficult task. Furthermore, the terms used, in general, had an equal distribution of both positive and negative correlations. This caused further difficulties in interpretation, but it can be assumed that the incorrect use of scales through the scoring of "double negatives" was a large contributory factor. "Double negatives" arose when assessors scored apparently "negative" terms (e.g. pale). Often, the scale was mistakenly used as a measure of the opposite "positive" term (e.g. dark) and hence a "double negative" effect occurred.

In an attempt to summarise the word usage, Tables 3, 4 and 5 were constructed and give a more detailed analysis of the term usage. The words and groups of similar words which correlated frequently with the axes were listed and counted. These frequencies were then calculated as a percentage of the total number of terms correlated with that axis. To find which terms were most frequently correlated with which axis, each row of percentages was summed. The use of the terms was then converted to percentages. The final column shows the percentages of term usage correlated with each axis. Using these tables it is now possible to interpret the consumer usage of the terms.

From Table 3, it can be seen that Axis 1 comprises 82% of the appearance terms that were correlated. A high correlation of Absorbance with Axis 1 had been

Table 3. Summary of terms correlated with Axis 1

Terms	a	b	c
maturity	12	5.15	28.97
smooth	11	4.72	22.36
mellow, velvety, rounded	0	0.00	0.00
mild, creamy, rich, soft	1	0.43	4.37
appearance terms (colour and depth)	108	46.35	82.02
flowery, aromatic, fruity, estery and perfumed	9	3.86	41.82
warm, fiery, burning	9	3.86	35.77
harsh, rough, sharp, coarse, nippy, bitter	17	7.30	15.95
malty	1	0.43	10.83
nutty	1	0.43	21.61
smoky, burnt wood, roasted	4	1.72	19.75
peaty, heathery, earthy	5	2.15	19.71
sweet, caramel, chocolate, treacle	9	3.86	25.23
	(187)		
others	46		
total	233		

The column headed a shows the frequency with which terms occurred; column b shows that frequency as a percentage of the axis total; and column c shows the percentage of the terms which were correlated with the axis.

obtained and it was concluded that Axis 1 described colour. Other groups of terms were also important for Axis 1. Maturity was quite highly correlated with Axis 1 and this demonstrates that a proportion of consumers relate perceived maturity with sample colour. Axis 1 has the lowest correlation of the three axes with the term Smoothness.

Axis 2 (Table 4) has a high correlation with Maturity. This Axis is shown to describe maturity effectively and can be used to determine which attributes the consumers associate with maturity. Peaty, Heathery, and Earthy feature strongly on the axis suggesting that these characteristics are linked with maturity. Smoothness is correlated more frequently with Axis 2 than either of the other axes.

Table 4. Summary of terms correlated with Axis 2

Terms	a	b	c
maturity	11	8.87	49.89
smooth	11	8.87	42.02
mellow, velvety, rounded	8	6.45	55.08
mild, creamy, rich, soft	7	5.65	57.42
appearance terms (colour and depth)	7	5.65	10.00
flowery, aromatic, fruity, estery and perfumed	2	1.61	17.40
warm, fiery, burning	3	2.42	22.43
harsh, rough, sharp, coarse, nippy, bitter	30	24.19	52.84
malty	2	1.61	40.55
nutty	1	0.81	40.70
smoky, burnt wood, roasted	4	3.23	37.08
peaty, heathery, earthy	9	7.26	66.54
sweet, caramel, chocolate, treacle	3	2.42	15.82
	(98)		
others	26		
total	124		

Column headings as in Table 3.

There are also a number of terms which relate highly to Axis 2 and also to Axis 3. On Axis 2 there is one group of terms which describe characteristics normally associated with immaturity. This group contains the terms Harsh, Rough, Sharp, Coarse, Bitter, and Nippy; these were largely negatively correlated to the axis. This negative correlation demonstrates that these terms are associated with the negative end of Axis 2 and hence appear to represent immaturity in the samples.

Axis 3 (Table 5) has the lowest correlation with Maturity, but many of the terms which had a high value on Axis 2 (and therefore were found to correspond to maturity) also have a high value on Axis 3. There is a high correlation of the terms describing Sweetness, Caramel, Chocolate and Treacle with Axis 3. Smoothness has its second highest correlation. It would appear that Axis 3 is describing a pleasant and desirable aspect of the

Table 5. Summary of terms correlated with Axis 3

Terms	a	b	c
maturity	5	3.76	21.15
smooth	10	7.52	35.62
mellow, velvety, rounded	7	5.26	44.92
mild, creamy, rich, soft	5	3.76	38.20
appearance terms (colour and depth)	6	4.51	7.98
flowery, aromatic, fruity, estery and perfumed	5	3.76	40.74
warm, fiery, burning	6	4.51	41.80
harsh, rough, sharp, coarse, nippy, bitter	19	14.29	31.21
malty	2	1.50	37.78
nutty	1	0.75	37.69
smoky, burnt wood, roasted	5	3.76	43.17
peaty, heathery, earthy	2	1.50	13.75
sweet, caramel, chocolate, treacle	12	9.02	58.95
	(85)		
others	48		
total	133		

Column headings as in Table 3.

whisky although these terms are less associated with Maturity and more with Smoothness and Sweetness.

The rating of Smoothness does not vary between axes in the same way as Maturity, but corresponds more closely to the group of terms Mellow, Velvety, Rounded and Mild, Creamy, Rich and Soft on all the axes. These terms would seem to describe factors contributing to smoothness and hence their values appear to influence the Smoothness rating. Maturity is apparently scored as an overall perception which encompasses many factors. Smoothness would seem to be one of the most important factors which contribute to the overall Maturity value without being directly associated.

This highlights the difficulty experienced in interpreting the axes. Almost all interpretations made have to be based on assumptions and many can be contradicted. Given this complexity, the explanation of word usage presented here is simplified.

This difficulty of interpreting the results of FCP has not been highlighted in the majority of published literature. Although FCP is claimed to reduce error by decreasing the influence of the panel leader and individual factors during the construction of a terms list, it would appear from this experiment that an even greater potential source of error is introduced during interpretation. Clearly defined meanings of the word lists are not available and so the interpretation will depend heavily on the researcher's own use of the terms. In other experiments the range of words generated and the variety of word usage have not been so large and hence the difficulties may not have been so apparent. This may be because previous panels used a small number of participants who were familiar with sensory techniques, were trained in the generation of sensory terms, and could use scales consistently.

Comparison of consumer and trained panel term usage

A comparison of the two panels' word usage was conducted and it was hoped to give a more informative, and less subjective, insight into the meaning of the consumers use of terms. The comparison, however, did not clarify the meaning behind the terms and the results remained difficult to interpret objectively.

Both panels have one axis which describes maturity and Axis 1 of the trained panel matches Axis 2 of the consumer panel in this respect. The high degree of agreement between the two sample plots demonstrates that the panels have used these axes in a similar manner. It can be inferred from this that both panels perceive maturity in the same way.

Consumer Axis 1 was found to represent some aroma notes; Axis 2 of the trained panel describes Woody and Floral notes. Woody and Floral were included in the aroma terms the consumers used and the panels appear to agree on the definitions of these characteristics.

On Axis 3, the panels did not have similar use of any terms.

CONCLUSIONS

The characteristics used by the consumer panel to distinguish between the whiskies in the set used here were Colour (Axis 1), Smoothness and Maturity (Axis 2), and Sweetness and Smoothness (Axis 3). Axis 2 appeared to describe the perceived characteristics of maturity and Axis 3 characteristics other than maturity which are perceived as pleasant. Consumers' perceptions of Maturity and Smoothness, or at least the terms they used to describe them, were very diverse. Neither Smoothness nor Maturity was readily described as another single characteristic.

The large variety of terms generated and the diversity of their usage made interpretation of the consumers' word use extremely difficult and laborious. Analysis of the term usage required many assumptions to be made and was highly subjective. Interpretation of word usage in consensus profiling is considerably simpler; there are clearly defined rules for this analysis and conclusions drawn are substantially less subjective.

The configuration obtained by FCP using consumers was very similar to the configuration generated by a trained consensus profiling panel. Therefore, with respect to the sample configuration, FCP by a panel of consumers can be said to be capable of producing results of equivalent quality to a trained panel using consensus profiling.

ACKNOWLEDGEMENTS

We would like to thank Chivas Brothers for supplying the samples; Gillian Arnold for the GPA treatment of data; Maggie Sheen for consumer testing advice, and of course all the assessors who made the work possible.

REFERENCES

Arnold, G.M., and Williams, A.A., (1986). The use of generalised procrustes analysis. In Statistical Procedures in Food Research (Piggott, J.R., ed.) Elsevier Applied Science, London.

Cairncross, S.E., and Sjostrom, L.B., (1950). Flavour profiles - a new approach to flavour problems. Food Technology **4**, 308.

Gower, J.C., (1975). Generalised procrustes analysis. Psychometrika **40**, 33-52.

Land, D.G., and Shepherd, R., (1984). Scaling and ranking methods. In Sensory Analysis of Foods (Piggott, J.R., ed.) Elsevier Applied Science, London.

Langron, S.P., (1983). The application of procrustes statistics to sensory profiling. In Sensory Quality in Foods and Beverages: Definition, Measurement and Control (Williams, A.A., and Atkin, R.K., eds.) Ellis Horwood, Chichester.

Piggott, J.R. and Canaway, P.R., (1981). Finding the word for it - methods and uses of descriptive sensory analysis. In Flavour '81 (Schreier, P., ed.) Walter de Gruyter, Berlin.

Piggott, J.R., Carey, R.G., and Canaway, P.R., (1985). Sensory analysis and evaluation of whisky. In Alcoholic Beverages (Birch, G.G., and Lindley, M.G., eds.) Elsevier Applied Science, London.
Reynolds, A.J., (1987). Sensory analysis by consumer. Food Manufacture, **Jan.**, 37-39.
Schutz, H.G., (1987). Predicting preference from sensory and analytical data. In Flavour Science and Technology (Martens, M., Dalen, G.A., and Russwurm, H.Jr., eds.) John Wiley & Sons, Chichester.
Stone, H., Sidel, J., Oliver, S., Woolsey, A., Singleton, R.C., (1974). Sensory evaluation by QDA. Food Technology **28**(11), 24.
Wierenga, B., (1986). Multi-dimensional models for the analysis of consumers perceptions and preference with respect to agricultural and food products, J. Agric. Economics 31, 83-97.
Williams, A.A., (1983). Defining sensory quality in foods and beverage. Chemistry and Industry 1983, 740-745.
Williams, A.A., and Arnold, G.M., (1985). A new approach to the sensory analysis of foods and beverages. In Progress in Flavour Research 1984 (Adda, J., ed.) Elsevier, Amsterdam.
Williams, A.A., and Langron, S.P., (1983). A new approach to sensory profile analysis. In Flavour of Distilled Beverages: Origin and Development (Piggott, J.R., ed.) Ellis Horwood, Chichester.
Williams, A.A., and Langron, S.P., (1984). The use of free-choice profiling for the evaluation of commercial ports. J. Sci. Food Agric. 35, 558-568.

5

Remotivating an industrial sensory panel

S. Marie
Food Science Division, Department of Bioscience and Biotechnology,
University of Strathclyde, 131 Albion Street, Glasgow, G1 1SD, Scotland

ABSTRACT

A distillery presented a problem to us arising because of continuing discrepancies between the results of flavour analyses of rectified spirit samples from the 'in-house' sensory panel and those of the customer's own panel. Poor motivation was diagnosed.

The problem was addressed with reference to the Hawthorne studies, reported by Roethlisberger and Dickson in 1939, where it was concluded that improvements in performance were due to the attention the workers received during a work-study exercise. After the same fashion, it was decided to attempt remotivation by attention.

Panellists agreed to be interviewed by a university researcher. Questions were designed to enquire about the history of each individual, both with the company and with the sensory panel, and to provide the opportunity for each panellist to both describe and comment upon all aspects of the sensory work.

Comparisons of panel performance, before and after the interviewing procedure, showed improved agreement between the two panels. In addition, the panellists' comments were illuminating and should serve as a guide for future management.

THE PROBLEM

A distillery presented a problem to us arising because of continuing discrepancies between the results of flavour analyses of rectified spirit samples from the 'in-house' sensory panel and those of the customer's own panel. The problem was thought by the management to lie mainly with the 'in-house' panel, since discrimination between samples was generally poor and because a negative attitude towards the sensory work was in evidence. Poor motivation was diagnosed.

TREATMENT BACKGROUND

The problem was addressed with reference to the Hawthorne studies, reported by Roethlisberger and Dickson in 1939. It was concluded by these authors that recorded improvements in performance of workers at an electrical company in Chicago were due to the attention the workers received during a work-study exercise, rather than due to any physical changes made to working conditions during the course of that exercise.

After the same fashion it was decided to attempt remotivation by attention, and an exercise was accordingly designed to involve the panellists in assessments of this aspect of their day-to-day activities.

TREATMENT GOALS

To obtain:

1. Closer agreement between the sensory analyses of the two panels.
2. General feedback to assist in the future management of the sensory work.

METHODS

Panellists

Twelve of the fifteen current panellists were available and agreed to be interviewed. Nine were office workers (typists, clerks and clerkesses, a telephonist, a supervisor and a manager) and three were laboratory staff (a chemist, and two laboratory assistants). Five were male and seven were female.

Interviewing procedure

The interviewer introduced herself as a researcher from the Food Science Department of the University of

Strathclyde, interested in industrial sensory panels and the experiences of the panellists in relation to the sensory work. She expressed a desire to present the results of this research at a conference, but in a general and statistical form so that no identities were revealed. Each interview took place in a private room, on a one-to-one basis, lasting approximately 20 min. Questions were structured, but open-ended, and panellists were encouraged to expand upon their replies and to comment fully.

The aftermath

The results of sensory analyses performed by these panellists were examined for periods immediately before (three months) and after (two months) the interviewing. Specifically, the frequencies of agreement between the two panels that the spirit was of a satisfactory standard (i.e. judged 'good') for each time period were compared using the chi-squared statistic.

Interview structure

Questions were asked on the following topics:

1. Personal details
   Name, position with the company, length of service and time on the sensory panel
2. The samples tested
   What were they? How much did they like them in general terms? How many were tested in a session? Any after-effects? Was the method of presentation acceptable, or could it be improved in their view?
3. Frequency of the sensory work
   How often was it done? Had there been changes in frequency over time? Were sessions regular and predictable, or spasmodic and unpredictable? Were there ever circumstances preventing their participation - if so, what were these?
4. Voluntary or compulsory status of the work
   Was their participation voluntary, or was it required?
5. Facilities and organisation
   What did they think of the facilities for the sensory work? Were the sessions well-organised? How could these be improved?
6. The importance of a panel leader
   How did they view the role of a panel leader? Would they like one now, as in the past?
7. Feedback
   Did they receive information or feedback about the samples they had tested? Through which

sources, if any?

8. The rewards
   Were the material rewards appropriate in content and frequency, in their view, and could they suggest an improvement (at little or no extra cost to their employer)? Were there other rewards besides the material ones?

RESULTS

Interview responses

The responses may be summarised as follows:

1. Five panellists had been on the panel since it started (approximately 8 years ago). The most recent arrival was 6 months ago, and the remainder reported 4, 5 or 6 years of service.
2. a) None of the panellists claimed to dislike the rectified spirits in general. Three said they liked them moderately, five liked them slightly, and four neither liked nor disliked them.
   b) There was general agreement that between 1 and 3 samples were nosed at any one session, exceptionally four.
   c) All panellists but two reported no after-effects whatever. The two exceptions mentioned a stinging, or pungent, sensation in the nose that has lasted for a very brief time (they seemed unperturbed by this).
3. a) Most panellists reported being called to do the nosing at least once daily, and 2 to 3 times daily in special circumstances. This pattern had not changed significantly over the years.
   b) The panel were split on their view concerning the regularity and predictability of the nosing sessions. Seven said they were regular and predictable, whilst four said they were spasmodic and unpredictable.
   c) Occasional irritation was expressed as a result of having to leave panellists' main tasks for the sensory work.
4. Ten panellists regarded the nosing work as required of them as part of the job, but six added that they didn't mind that. Two thought it voluntary.
5. a) Ten of the twelve panellists considered there were problems with sample presentation and sensory facilities. These were mainly concerned with smells that masked or competed with the spirit aromas and made the nosing work more difficult. External processing smells were mentioned, as well as internal smells such as cigarette smoke from an adjacent room, perfume,

aftershave and spilled whisky. It was pointed out that such internal smells were not readily dissipated because of the small size of the nosing room, despite the operation of an extractor fan. Crowding, and opportunities for collusion between panellists, because of the small, communal area, was mentioned.
b) Nine panellists recognised organisational problems of having to co-ordinate the sensory work with the arrivals of tankers, and thought these problems inevitable. However, it was suggested that communications for calling panellists might be improved.

6. Only one panellist saw a need for re-introducing a panel leader. Two panellists thought the role of a panel leader unimportant. Four said things were fine as they were and regarded the laboratory assistant who presently put out the samples as a panel leader.
7. Five panellists said they received no information or feedback with regard to the samples they had tested. Four obtained feedback solely as a result of the nature of their own jobs and thought information should be more generally available.
8. a) Five panellists said they were happy with 'bottles' as rewards, and five said alternatives were worth considering, such as larger but less frequent gifts, more personal gifts (bottle contents were not always personally consumed), and a voucher system whereby each session attracted a voucher that could be saved up and redeemed for a choice of gifts. One panellist thought it was time for an improvement in the rewards, and suggested more bottles or more 'premium' bottles.
b) Of the suggested sources of non-material reward, the fact that the panel could not presently be replaced by a machine aroused some interest. Three appreciated being involved with the final product, six found the work intrinsically interesting, whilst four did not (one said because of insufficient knowledge), and four said they recognised the worth of the panel to the company.

Agreement between panels

Agreement frequencies between the two panels for the periods before and after the interviewing are shown in Table 1.

Table 1. Frequency of agreement between the two panels

	Agreement Yes	Agreement No
Before Interview	10	27
After Interview	10	10

The frequency of agreement was, thus, significantly higher after the interviewing (chi-square = 3.00 with 1 df; $p < 0.10$).

THE OUTCOME

Goal 1

There was some preliminary evidence that the company's own 'in-house' panel was performing more in line with the customer's panel after the interviewing exercise. Although the discriminatory proficiency of the customer's panel is unknown, quality control must necessarily be geared to the customer.

Goal 2

Assuming that this improved performance was a real effect, and that self-esteem and motivation were increased, it is important that these benefits are sustained in the interests of continuing compatibility of the two panels. Such Compatibility depends, of course, on the stability of the customer's panel, also. The comments made by interviewed panellists should serve as a guide to the means of such sustenance. The most important points that emerged may be summarised as:

1. The physical conditions.
   Some arrangement that excludes external and internal smells from the nosing area is most desirable, since nearly all panellists commented upon this problem. Individual booths would minimise the problem of crowding and collusion between panellists.

Management's response: A new nosing room is planned.

2. Feedback.
   There was a consistent call from the panellists for information, both of an educational nature to promote panellists' interest in the work and for feedback about the results of testing i.e. acceptance or rejection by the customer.

Management's response: Initially, results will be posted on the wall of the nosing room. More information will be included in the monthly meetings attended by all staff.

3. The panel leader.
   This may not be the problem that was supposed by the management since panellists seemed content with the present situation. However, the allocation of more time by a person in such a role would evidently be appreciated by the panel for the purposes of more training and as an information source.

Management's response: There will be a re-allocation of time for the present organiser in favour of the sensory work.

4. Rewards.
   Half of the panel were content with the reward system as it stands whilst half were interested in alternatives. A voucher system could be a powerful motivator, and is based on the same principles as the 'token economy' used by clinical psychologists. The giving of a voucher on each testing occasion would more firmly associate the work with the reward, and the incentive to work for a chosen item (for which vouchers could be saved and redeemed) would be more dynamic and reinforcing than the almost passive acceptance of a regular monthly bottle.

Management's response: This matter is under consideration.

REFERENCE

Roethlisberger, F.J. and Dickson, W.J. (1939). Management and the Worker. Harvard University Press.

6

Computerised collection and statistical analysis of descriptive sensory profiling data

M. L. Arnott
Reading Scientific Services Ltd., The Lord Zuckerman Research Centre, The University, Reading, RG6 2LA, UK

ABSTRACT

A computerised sensory analysis system is used to collect and analyse the data arising from descriptive sensory profiling of products. The sensory booths are equipped with microcomputer keyboards and monitors through which assessors input their data. These microcomputers are networked and controlled by a file server/master microcomputer which also handles collation of data. A statistical package is used for the routine analysis of the sensory profiling data which normally comprises analyses of variance and principal components analysis. The system can also be used to run discrimination tests, category scaling, free choice profiling and time intensity studies.

Descriptive profiling can be used to compare finished products or monitor flavour development during processing; examples are discussed.

INTRODUCTION

In recent years considerable advances in sensory evaluation techniques have been achieved by the use of computers and particularly microcomputers. Microcomputers can be used to prepare sensory questionnaires, enter and collect data, perform statistical analysis, assist in panel training and in the production of reports and graphics [1,2]. Computer systems have been developed specifically for use in sensory laboratories [3,4,5] and these vary in their complexity and flexibility. When sensory methods such as descriptive profiling are used routinely large amounts of data are generated and the use of a computer system which can collect and analyse such data is essential.

DESCRIPTIVE SENSORY PROFILING

The technique of quantitative descriptive analysis, (QDA) [6], is used to describe the sensory characteristics of a product and to quantify the key characteristics which discriminate between samples. This profiling method can be applied to any type of product, detergents, perfumes, polishes, fabric, tobacco as well as food and drink and also to the specification of processes and raw materials.

A screening procedure is used to select QDA assessors who then undergo a training programme of lectures, demonstrations and panel sessions. Different types of product are examined and the trainees become familiar with the terms used to describe sensory attributes. At the end of training each individual should be able to perceive and recognise individual sensory characteristics and with the use of appropriate standards reach an agreement with fellow assessors on the descriptors to be used and assess their intensities.

QDA evaluation of a particular product type starts with the assessors developing an appropriate vocabulary of odour, flavour, aftertaste and mouthfeel or texture terms. This development of a fixed product vocabulary is a crucial stage of the QDA technique. The vocabulary is used by the assessor to differentiate between products and if it is inadequate this will not be possible. If a product vocabulary has been established previously it is permissible to familiarise the assessors with the products, descriptors and definitions, including odour standards whenever possible. However, the assessors must be satisfied with the terms before proceeding to a product evaluation. The identification of a new perception or a request to change a descriptor or a definition is the collective responsibility of the panel and if necessary the panel manager must ask the panel to reconvene for discussion. The assessors must be regularly reminded of this responsibility which allows for some flexibility and ensures that the current vocabulary does not become outdated.

During an evaluation the assessors assign

individually an intensity to each term for each sample. The assessments are performed at least in duplicate, as randomly presented samples identified only by code numbers. The assessors mark their perceived intensity for each attribute on an open line with word anchor points at each end of the line, the intensity scale always increasing from left to right; absent to extreme, weak to strong, etc. By measuring the distance from the left hand anchor point to the perceived intensity mark a numerical value is obtained for analysis.

As a single sensory experiment can use a product vocabulary of over 50 terms and involve a panel of eight or more assessors evaluating multiple samples on more than one occasion, it is normally the case that considerable quantities of data are generated. Originally assessors recorded their perceptions on paper sheets pre-printed with a line scale for each descriptive term. The pencil marks on the line scales were later converted into numerical scores and entered by hand onto an IBM PC for statistical analysis. This was obviously a laborious and time consuming operation in urgent need of automation.

DATA COLLECTION

The manual system used originally has now been replaced by a computerised data input and collection system. Each of the sensory booths is equipped with a microcomputer, keyboard and monitor through which assessors input their data. These microcomputers are networked and controlled by a file server/master microcomputer which also handles collation of data. Our present system uses BBC microcomputers for the input and collection of data with software originally developed at the Agricultural and Food Research Council at the Institute of Food Research, Bristol. A new software package "TASTE" [7] has been written which will run a system based on IBM PC or IBM PC-compatible microcomputers. This system offers certain advantages in terms of hardware options and greater flexibility in setting up experimental details for panel sessions. Both versions are menu driven so that the sensory analyst setting up the experiment, and the assessor evaluating samples, react to simple sequential questions. With the BBC system all data must be entered using the microcomputer keyboard. Descriptive profiling data are entered by manipulating the cursor keys, but for other tests assessors can type appropriate responses. In practice, this has not presented a problem and assessors soon become familiar with the use of the keyboard and respond positively to the introduction of microcomputers in the booths. There are several options available for the input of data with the IBM PC based system. A keyboard can, of course, be used but alternatives are the use of a mouse or a smaller compact keyboard with a reduced number of functions. The method of choice would be determined by the type of sensory tests to be carried out and the space available to adapt the booths. The type of VDU screen can also be varied and, when space is at a premium, smaller flat screens are available. The operation of the system is basically the same for both versions.

A sensory project can be designed and set up well in advance of the panel sessions using the menu driven system. The main menu offers seven options:

1) Specify Task Name
2) Parameter File Handling
3) Question File Handling
4) Set up a Panel Session
5) Print Panel Results
6) File Utilities
0) Finish

The task name is the identification for the current project and all previous project names can be recalled if necessary. The system is safeguarded against the use of the same task name for more than one project.

The Parameter File holds all the information relating to the design of the experiment such as title, details and numbers of test samples, number of assessors and sessions. A new file for a specific experiment can be created or an existing parameter file modified. There are sub-menus to enter the number of panel sessions, number and identification of assessors. The sample details can include the overall number of samples, the names of the samples, the numbers of samples in each session, the sample codes and the presentation order.

The Question File specifies the type of sensory test that is required and the current options are descriptive profiling (QDA), category scaling, difference tests, free choice profiling and time intensity studies. Within each project more than one question may be set, for instance a combination of descriptive profiling and category scaling could be used where the category question could be used to identify a particular attribute type. Within the descriptive profiling vocabulary there may be an attribute such as nutty. The assessor may be asked to score for overall intensity of nutty and then asked to identify the nut type, e.g. hazel, walnut, peanut. As with the Parameters File the Question File can either be set up as a new file or an existing file modified. This is particularly useful in descriptive profiling when similar vocabularies are to be used with the addition or deletion of only a few descriptors. Up to ten attribute categories such as odour, flavour, mouthfeel, each consisting of up to ten attribute terms can be set up. An attribute category consisting of ten terms with associated line scales can be displayed on the screen at one time. The anchor words associated with the line scales can be different for each attribute if required.

Once the project details have been entered and Parameter and Question Files established the panel session can be set up by specifying the task name and the panel session required. During the panel session the microcomputer screen in the booth displays instructions and presents the pre-defined vocabulary for scoring. The scores are entered using the cursor keys and the screen displays are advanced as required using the return key or space bar as instructed.

When the assessors have completed the panel session their data are stored awaiting completion of further sessions in the project or analysis.

The File Utilities provides functions for converting the panel data to files which can be accessed for statistical analysis and holds all the assessor information, names, addresses, previous panel experience.

DATA ANALYSIS

On completion of the whole experiment the data are analysed using SENPAK, a statistical package specially designed for the analysis of sensory profiling data [8]. The prime objective of statistical analysis is to determine whether there is a difference between samples but it can also be used to monitor assessor performance. The consistency of individual assessors within experiments and of the panel from experiment to experiment can be investigated. The need for further training, either of specific individuals or of the whole panel, is assessed by statistical measures such as mean, standard deviation, rank order and analysis of variance.

The sensory data are analysed using analysis of variance showing both assessor and product sample effects. Assessor-sample interactions are also estimated when the experiment includes replicates. The analysis of variance enables the sensory analyst to determine whether the mean scores for several samples differ from one another in sufficient magnitude to justify considering them discriminated at some stated level of significance. This analysis does not specify which samples are different so some additional computation must be carried out using a multiple comparison test. Several such tests, Tukey, Newman-Keuls, Duncan and Fisher's Least Significant Difference (LSD) tests can be used to determine those pairs of sample means that are significantly different. All the procedures involve calculating a range that is matched to the difference between the means. When differences between mean values are greater than the range value, the samples are significantly different. The choice of multiple comparison test will vary with the circumstances of the experiment, but when few comparisons need to be made the LSD is most likely to find significant difference. Caution should be exercised in the interpretation of results and no one statistical measure will be totally satisfactory.

At this stage in the data analysis the sensory analyst can draw conclusions and present results. It is often helpful if such results can be presented in an easily understood and compact visual format. This could be achieved simply by plotting sample mean scores in the form of histograms by importing SENPAK files into a program that will produce graphs such as LOTUS 1-2-3. If only two or three samples are being compared a star diagram plot may be a better illustration of the differences between samples. The intensity values for the various attributes are plotted on a series of lines that radiate from a centre point. The relative intensity for each attribute increases with distance from the centre point. Often only the attributes which have been identified by analysis of variance to show statistically significant differences between samples are plotted. The function for star diagram plots is an integral option in the SENPAK program with the facility to rescale the attribute intensities to produce clear plots. The example of a star diagram (Figure 1) shows the characteristics of two home brew beers when different yeast cultures were used for the fermentation process. All the attributes plotted were shown to be significantly different by analysis of variance. Figure 1 shows that the normal dried pitching yeast, which contained a mixed flora, produced a beer with a more intense yeasty, caramel, sweet, hoppy, fruity flavour and a more nutty flavour and sweet aftertaste with more dissolved gas giving a higher carbonated score. The experimental pure culture yeast gave a beer with a more sulphury odour and flavour and aftertaste, a more soapy flavour,

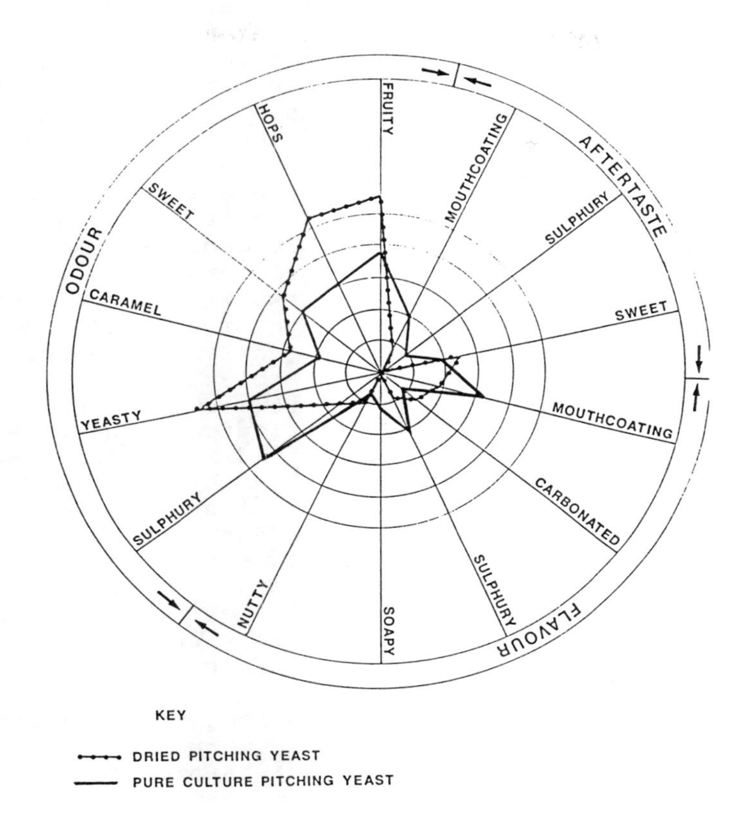

Figure 1. Sensory profile analysis of pure culture beer fermentation

less gas and a mouthcoating sensation which carried over into the aftertaste. This particular yeast was obviously not the strain to include with the home brew kit!

Data can also be presented in the form of a principal components analysis (PCA) plot which is also an integral function of the SENPAK program.

PCA is a multivariate analysis technique which determines the relative positions of samples in multi-dimensional space. It reduces a set of individual items into components such that the first component has maximum correlation with all the variables and accounts for the greatest amount of variance, the second component for the second largest amount of variance and so on until as much of the variance has been accounted for as is reasonable. In practice it is usual to find that the first and second principal components together describe a substantial proportion (over 75%) of the total variability present in the sample population. This gives considerable confidence in the validity of the relationships illustrated by a two-dimensional diagram in which the first and second principal components are shown as axes at right angles to one another. It is not usually necessary to proceed beyond the second principal component. In addition to locating samples in a two-dimensional format PCA calculates values (loadings) for each sensory attribute included in the analysis and uses these loadings to determine the relationship of each attribute to each other attribute and to the samples.

Normally only those attributes shown by ANOVA to contribute to a significant difference between samples would be included in the PCA. A PCA diagram, therefore, illustrates both the relative positions of samples to one another and the attributes which differentiate them. The further an attribute is located from the origin along a principal component, the more important it is in defining that principal component. The closer an attribute is to the position of a sample the more characteristic is the attribute of that sample.

The use of descriptive sensory profiling coupled with statistical analysis can provide the food technologist with important information which can help identify key points in flavour development during processing. The PCA plot (Figure 2) shows the effects of drying sweetened milk powders for varying periods at a temperature of 80 °C. It can be seen that a very short period of drying resulted in a number of undesirable flavour notes such as rancid, rubbery and cheesy. A period of 2-4 hr produced a product with creamy characteristics while periods in excess of this gave rise to malty, caramelised and buttery characteristics. If for instance the objective was to produce a product with creamy characteristics then a 2 hr process could be used as this was not significantly different from a 4 hr process in terms of flavour development but could result in considerable cost savings.

In summary, it can be seen that the technique of descriptive profiling facilitated by computerised data collection and analysis can provide a powerful tool which can be used with benefit in process development, product development and finished product analysis.

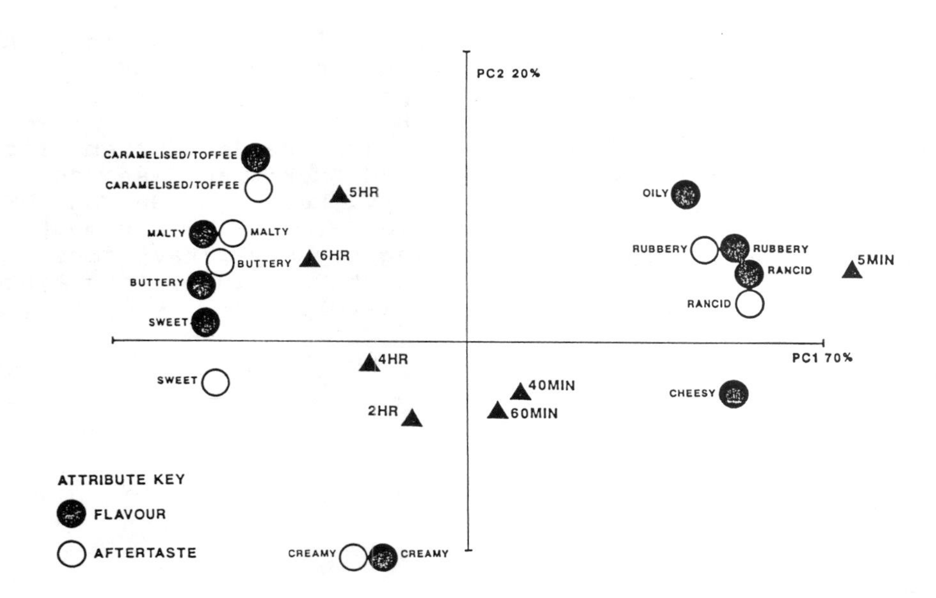

Figure 2. Principal components analysis of sweetened milk powders with varying drying times

Our own experience of computerising the sensory laboratory is that it produces no insoluble problems for panellists but raises substantially the efficiency of data capture and analysis. The saving in staff time alone is sufficient to effectively repay the capital outlay for the system in less than one year. In the future computerised sensory laboratories will be the norm and not the exception.

REFERENCES

1. Williams, A.A., Brain, P. The scope of the microcomputer in sensory analysis. Chemistry and Industry, 1986, 4, 118-122.
2. Walchak, C.G. The use of the computer in the sensory evaluation laboratory. Food Tech. in Australia, 1986, 38(4), 152-154.
3. Findlay, C.J., Gullet, E.A., Genner, D. Integrated computerised sensory analysis. Journal of Sensory Studies, 1986, 1, 307-314.
4. Lyon, D.H. Sensory analysis by computer. Part I. Food Manufacture, 1986, Nov., 40-42.
5. Piggott, J.R. Automated data collection in sensory analysis. In Flavour of distilled beverages: Origin and development (Piggott, J.R., ed.). Ellis Horwood Ltd., Chichester, 1983.

6. Stone, H., Sidel, J., Oliver, S., Woolsey, A., Singleton, R.C. Sensory evaluation by quantitative descriptive analysis. Food Technology, 1974, **28**(11), 24.
7. "TASTE" Software Package. Reading Scientific Services Ltd, The Lord Zuckerman Research Centre, The University, Reading, RG6 2LA. UK.
8. Williams, A.A., McFie, H. SENPAK: A computer program for analysing sensory data. Proceedings of the 9th Wine Subject Day, 1985, Long Ashton Research Station, Bristol.

7

Chromatography with supercritical fluids

Mark W. Raynor, Jacob P. Kithinji, Ilona L. Davies and Keith D. Bartle
Department of Physical Chemistry, University of Leeds, Leeds, LS2 9JT, UK

INTRODUCTION

Chromatography with supercritical fluids was reported more than twenty years ago [1]. However, its advantages have only been fully realised recently and these are especially relevant to the analysis of involatile, thermally unstable and polar compounds [2]. Capillary gas chromatography (GC) for example has the highest resolving power of the chromatographic techniques, but is limited by the volatility and thermal stability of the sample. High performance liquid chromatography (HPLC) although able to analyse polar, unstable and involatile compounds has limitations because of the low efficiences associated with packed columns and because there is still no simple universal detector available for detecting compounds which do not have a chromophore. Supercritical fluid chromatography (SFC) is able to perform high resolution separations of these types of compounds at temperatures well below those of thermal decomposition [3]. The technique can also be interfaced with almost any HPLC or GC detection system.

This chapter introduces SFC as an analytical technique which could have significant application in flavour research. Some of the basic principals and instrumental aspects of SFC are discussed together with several examples (such as the analysis of aromatic acids) where the technique could be of particular use.

PRINCIPALS

The basic principal of SFC as in GC and HPLC is partition between the mobile phase and the stationary phase, which is either a packing of silica particles on to which are bonded a variety of functional groups (as in HPLC columns), or a crosslinked film of elastomeric stationary phase coated on the inner wall of a capillary column (as in GC) [4].

The supercritical mobile phase may be characterised from the phase diagram of a substance in Figure 1. The solid, liquid and gaseous states are well defined, but above the critical temperature and pressure, the meniscus between the gaseous and liquid states dissappears. In this region the density and solvating power of the fluid approaches that of a liquid, but its viscosity is similar to that of a gas and its solute diffusion characteristics are intermediate between the two. Table 1 compares the three mobile phases with regard to these properties.

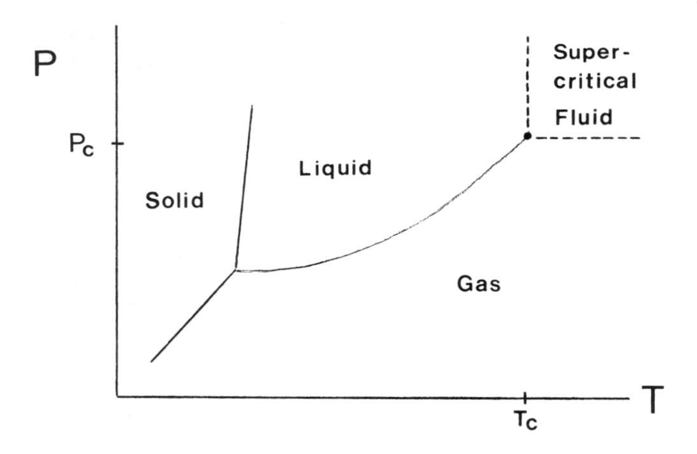

Figure 1. Phase diagram of a pure substance.

Table 1. Comparison of mobile phases for GC, SFC and HPLC.

Mobile phase	Density (g/ml)	Viscosity (p)	Diffusion Coef. (cm /s)	Column type
GC	10^{-3}	$0.5\text{-}3.5 \times 10^{-4}$	0.001-1.0	capillary
SFC	0.2-0.9	$0.2\text{-}1.0 \times 10^{-3}$	$3.3\text{-}0.1 \times 10^{-4}$	packed & capillary
HPLC	0.8-1.0	$0.3\text{-}2.4 \times 10^{-2}$	$0.5\text{-}2.0 \times 10^{-5}$	packed

Under supercritical conditions, fluids possess properties which are highly desirable for chromatographic mobile phases: high solvating power allows the migration of high molecular weight material; low viscosity results in low pressure drops across the column; high diffusivities of solutes in the mobile phase confers excellent mass-transfer properties so that higher efficiencies per unit time are achieved than are possible with HPLC. In addition, solute retention in SFC is influenced by the density of the mobile phase. Hence, pressure or density programmed elution in SFC is analagous to temperature programming in GC and gradient elution in HPLC.

Potentially useful mobile phases for SFC must possess moderate critical parameters and have good solvating power under supercritical conditions [5]. Some of the pure mobile phases which have been investigated for SFC are shown in Table 2. Carbon dioxide is most commonly used because it has critical parameters which are easy to work with (T_c = 31.3°C , P_c = 72.9 atm) and it is non-toxic and non-inflammable. The addition of small quantities of polar solvents to the mobile phase as modifiers can greatly alter the retention behavior of analytes and so modified mobile phases are being increasingly used for separating polar compounds [6,7]. Other mobile phases such as xenon have received attention because of specially favourable properties such as detector compatibility.

Table 2. Pure mobile phases for SFC

Fluid	Critical Temperature (°C)	Critical Pressure (atm)	Dipole moment
CO_2	31.3	72.9	0
SF_6	45.5	37.1	0
Xe	16.6	57.6	0
n-C_5H_{12}	196.6	33.3	0
$CClF_3$	28.8	39.0	0.50
N_2O	36.5	71.4	0.51
NH_3	132.4	111.3	1.65

INSTRUMENTATION

A block diagram of the instrumentation required for SFC is shown in Figure 2. The mobile phase (see Table 2) is drawn from a cylinder as a liquid and pressurised in this state by a syringe pump before being delivered via an injection valve to the analytical column. The column is installed in an oven which is heated above the critical temperature of the mobile phase and is connected to a pressure restrictor which restricts the mobile phase flow. The restrictor is either installed within the detector, such as the ion source of a mass spectrometer (MS) and the flame tip of a flame ionization detector (FID) or it is placed in-line after flowcell-type detectors such as the ultraviolet absorbance (UV) and fluorescence detectors.

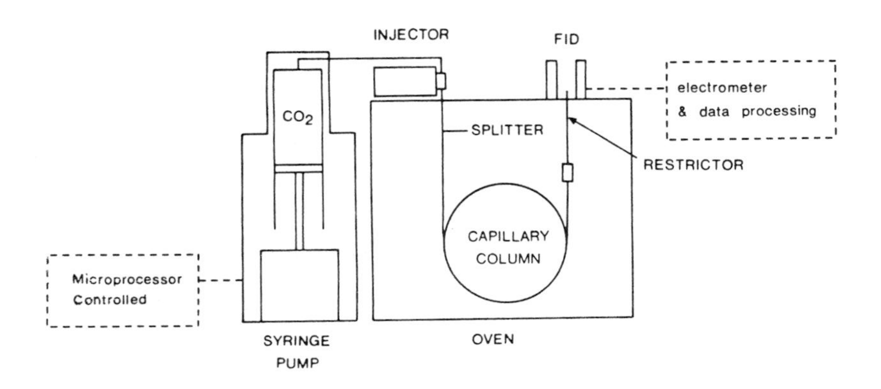

Figure 2. Instrumentation for SFC

The mobile phase must be selected largely on the application and detection requirements. Flame based detectors are only compatible with fluids such as carbon dioxide, or nitrous oxide which have a low background signal in the flame. The purity of the fluid must also be controled, otherwise, on programming the density, a rise in baseline can occur. Other common detectors such as the UV detector can be used with a wider variety of mobile phases which are transparent in this region of the electromagnetic spectrum, such as pentane, carbon dioxide modified with methanol etc.

Pulse free syringe pumps are used for SFC (particularly capillary SFC) because of the need to deliver the mobile phase at a constant pressure rather than flow rate. Since the solvating power of the mobile phase is proportional to the density of the fluid, the pressure is usually manipulated at constant temperature for density control [8]. This is performed by using a sensitive pressure transducer in-line between the pump outlet and the column, and a microprocessor for pump control. The microprocessor is programmed to generate a set pressure profile.

The method of sample introduction into the column is important for preserving chromatographic efficiency. For packed column SFC, a standard high-pressure HPLC injection valve is employed with a 5 or 10 µl sample loop. In the case of capillary SFC a submicrolitre injection valve must be used because of the low sample capacity of the column. Small sample volumes are injected either by rapidly switching the valve from the load to inject position and back again (time split injection) or by using an inlet splitter.

The heart of the system however is where the separation takes place, and in this respect SFC has actually developed along two complimentary lines that differ in nature of the column used [9]. Early work in SFC was performed on packed columns using adsorbants such as alumina or silica. Developments in this area have resulted in bonded non-extractable stationary phases such as octadecyl and cyanopropyl bonded silicas, which have different selectivities. Packed columns are advantageous when the main objective is to isolate individual analytes quickly. The sample capacity of these columns is much greater than capillary columns and hence larger amounts can be injected on to the column before overloading occurs. Packed columns also have more theoretical plates per unit time than capillary columns, but do suffur from interactions with reactive solutes.

Consequently wall coated capillary columns with diameters less than 100 µm are being increasingly used in SFC. Stationary phases for capillary SFC are based on a polysiloxane backbone, crosslinked by means of a free radical initiator and have incorporated into them phenyl, biphenyl, octyl and cyanopropyl substituents for selectivity characteristics. These columns permit maximum use of density programming and offer higher efficiencies than packed columns because of greater length (±10 m) consequent on the permeability of capillaries to the flow of low-viscosity supercritical fluids. Their low flow rates allow the use of toxic, corrosive or expensive mobile phases and also make them compatible with a variety of detectors.

The ease with which SFC can be interfaced with both HPLC & GC detectors has been one of the major reasons for its rapid development [10,11]. GC flame based detectors such as the flame ionization detector (FID), flame photometric detector, nitrogen phosphorous thermionic detector and electron capture detector have all been interfaced with SFC systems using a capillary restrictor which effectively decompresses the fluid to ambient pressure just as it enters the flame tip. UV and fluorescent detectors are more commonly used with packed column systems operating with organic or modified mobile phases. Of course detectors which provide structural information about the analyte such as mass spectrometry (MS) and Fourier Transform infrared spectrometry (FTIR) are the most useful detectors for qualitative analysis. Both of these detection systems have been successfully coupled with SFC because of the gaseous nature of the mobile phase on decompression: the column effluent may be passed directly into the ion source of the MS; or collected at a solvent elimination interface for subsequent FTIR analysis.

APPLICATION TO DISTILLED BEVERAGE FLAVOURS

SFC has already been applied to a variety of samples including drugs, foods, natural products, explosives, biological and fossil fuel materials and polymers [2,5]. The number of applications continues to grow and potentially distilled beverage flavours is another area where SFC could be of use. Although the more volatile components in these flavours are more amenable to GC analysis, there are a significant number of compounds which may not be detected because of their involatility and/or polarity. Aromatic acids (such as vanillic acid for example) fall into this category. In order to separate these compounds by GC, they must first be derivatised to their more volatile and less polar esters. This is time consuming and often impractical when dealing with complex mixtures. HPLC methods on the other hand are often unable to achieve the required chromatographic resolution even with complex eluent gradients. Results which are currently achievable with capillary SFC are shown in Figure 3.

Fifteen aromatic acids from benzoic acid to trans-p-hydroxycinnamic acid have been separated on a 10 m x 50 µm capillary column coated with an oligo(ethylene oxide)-substituted polysiloxane stationary phase [12]. Carbon dioxide was used as the mobile phase and was density programmed at 110°C from 0.5 g/ml to 0.74 g/ml at 0.015 g/ml/min after an initial 5 minute isoconfertic period. Detection was by FID at 400°C.

These acids are highly adsorptive solutes, but are eluted from the capillary column as sharp peaks with little tailing. This is due to the use of a well deactivated capillary column and the solvating characterisitics of the carbon dioxide. The acids could not be eluted from a packed column SFC with carbon dioxide due to activated sites in the column packings. Over the duration of this program, the density of the mobile phase, and hence its solvating power, increases significantly and this enables progressively more polar acids to be eluted. There is however a density at which the solvating power of the mobile phase reaches a limit. At this point the peaks begin to broaden: an effect which is analagous to the elution of involatile compounds at the final temperature of a GC temperature program. A more polar mobile phase would therefore have to be employed to elute di- & tri-hydroxylated aromatic acids.

Other groups of compounds which are present in distilled beverages are the phenols and aromatic aldehydes. These solutes can also be separated by capillary SFC as shown in Figure 4. This separation was carried out on the same column as the aromatic acids. In this instance, the mobile phase (carbon dioxide) was pressure programmed from 80 atm to 350 atm at 4 atm/min after an initial 8 minute isobaric period. The column was held at 100°C during the analysis.

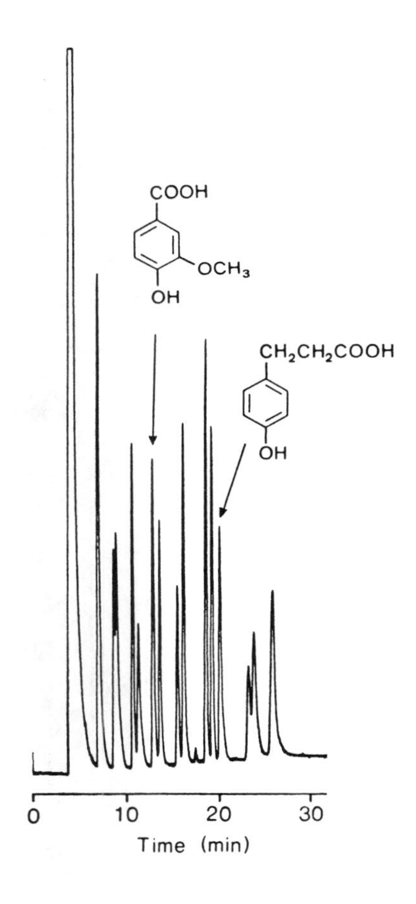

Figure 3. Separation of aromatic acids. See text for chromatographic conditions. Elution order: benzoic acid; cinnamic acid; 4-chloro-benzoic acid; 3,5-dimethoxybenzoic acid; chlorocinnamic acid; vanillic acid; 3,4-dimethoxycinnamic acid; syringic acid; ferrulic acid; sinapic acid; p-hydroxyphenylpropionic acid; p-hydroxyphenyl-acetic acid; p-hydroxybenzoic acid; cis-p-hydroxycinnamic acid; trans-p-hydroxycinnamic acid.

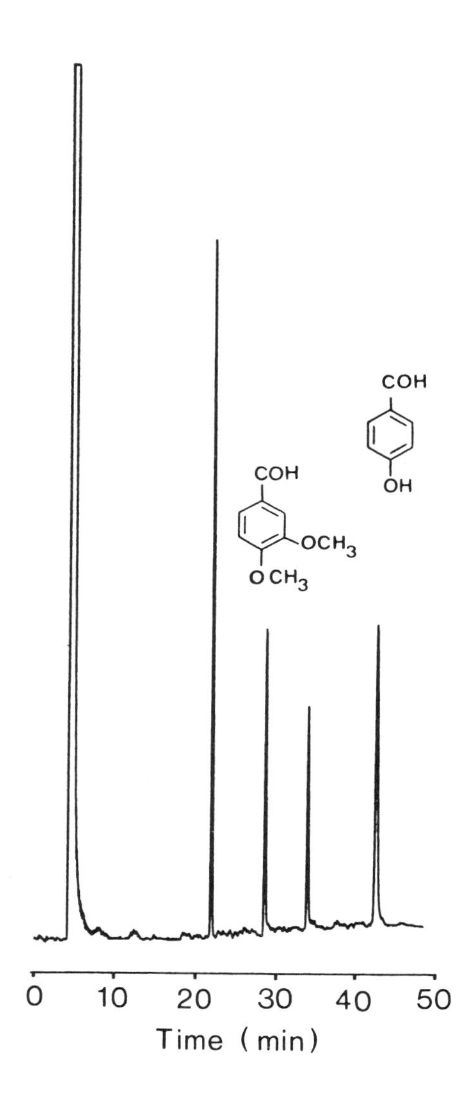

Figure 4. Capillary SFC separation of 4-allyl-2-methoxyphenol, vanillin, syringealdehyde and p-hydroxybenzaldehyde (elution order as written). See text for chromatographic conditions.

CONCLUSIONS

The future of SFC depends largely on its ability to help the analyst solve difficult analytical problems. The technique has definite advantages over HPLC and GC for certain applications. In addition, when combined with MS, SFC could be an invaluble tool for separating & identifying compounds which are not amenable to analysis by GC/MS. The preliminary results shown in this chapter demonstrate its potential use in flavour analysis.

ACKNOWLEDGEMENTS

The authors thank T. Katase and J. Piggott for supplying the acid, phenol and aldehyde samples, C. Rouse for preparing the glyme capillary column, and the Science & Engineering Research Council (grant number GR/E/00556) for financial support.

REFERENCES

1. E. Klesper, A.H. Corwin & D.A. Turner, J. Org. Chem. 27 (1962) 700.
2. C.M. White & R.K. Houck, HRC&CC 9 (1986) 3.
3. T.L. Chester, J. Chromatgr. Sci. 24 (1986) 226.
4. P.A. Peadon & M.L. Lee, J. Liq. Chromatogr. 5 (suppl. 2) (1982) 179.
5. D.W. Later, B.E. Richter & M.R. Anderson, LC-GC 4 (1986) 993.
6. B.W. Wright & R.D. Smith, J. Chromatogr. 355 (1986) 367.
7. M.W. Raynor, J.P, Kithinji, I.K. Barker, K.D. Bartle & I.D. Wilson, J. Chromatog. 436 (1988) 497-502.
8. M.L. Lee & K.E. Markides, Science 235 (1987) 1342.
9. P.J. Schoenmakers & L.G.M. Uunk, European Chromatogr. News, 1 (1987) 14.
10. M. Novotny, HRC&CC 9 (1986) 137.
11. D.W. Later, D.J. Bornhop, E.D. Lee, J.D. Henion & R.C. Wieboldt, LC-GC 5 (1987) 804.
12. C.A. Rouse, A.C. Finlinson, B.J. Tarbet, J.C. Pixton, N.M. Djordjevic, K.E. Markides, J.S. Bradshaw & M.L. Lee, Anal. Chem. 60 (1988) 901.

8

Measurement of food and beverage colour appearance

D. B. MacDougall
AFRC Institute of Food Research, Bristol Laboratory, Langford, Bristol, BS18 7DY, UK

SUMMARY

Colour is a psychological experience. Systems for measuring colour are based on the well established relationship that human vision has a tri-receptor mechanism which adapts to the surroundings and the light. The development of the CIE uniform colour scales is described and the limitations of applying standard colour specification to the appearance of translucent and transparent foods discussed. Specifically, data on the absorption spectra of distilled alcoholic beverages, with and without added caramel, are presented and related to their lightness, hue and chroma and the effects of change in lighting quality on their hue and chroma calculated.

INTRODUCTION

Acceptance of foods and beverages at retail depends on the consumer's experience of the product, its cost and its appearance. The most important aspect of food appearance is colour, especially if it is related to other aspects of quality. Rejection by the consumer at point of sale is usually the consequence of pigmentation and structural faults in the food which may so alter its appearance that it is assumed to be inferior or spoiled.

APPEARANCE MEASUREMENT

Colour can be regarded as the psychological interpretation of the information in the light reflected, transmitted or scattered by an object as influenced by the environment in which it is presented. It is controlled by a variety of interacting factors: observer memory, adaptation and colour vision defects, such as red-green colour blindness. The illumination variables of colour temperature, intensity and rendering all influence the visual processing of the colour stimuli. Colour, however, is only one component of appearance. The physical variables of translucency, opacity and gloss are equally important in the appreciation of appearance and must be accounted for if descriptions are to be meaningful and measurement procedures standardised. This is accomplished in two stages, the physical and the psychological. This first includes measurement of the object's dimensions, its structure, pigmentation, location in space and relationship to other objects in view. The second is accomplished by translating the object's reflectance or transmittance spectrum into colour space by psychophysical constants derived from human colour vision response and the spectra of light.

Colour terms can be divided into the subjective and objective (Hunt, 1978). The subjective are brightness, lightness, hue, saturation, chroma and colourfulness. Colourfulness is that aspect of visual sensation according to which an area appears to exhibit more or less chromatic colour. Hue is that attribute described in colour names (red, green, purple, etc.) but saturation and chroma are less easily comprehended. Saturation is colourfulness judged in proportion to its brightness whereas chroma is colourfulness relative to the brightness of its surroundings. A similar difficulty exists in understanding the difference between lightness and brightness. Lightness is relative brightness and is unaffected by illumination level whereas brightness increases with increase in illumination.

Foods exhibit every possible appearance characteristic. They may be wet, dry, shiny, diffuse, irregular or flat. Their structure causes them to appear transparent, translucent or opaque and the amount and type of pigmentation creates the stimuli for colour which may be uniform, patchy or graded. Procedures recommended for measuring flat opaque colours such as paint and tiles may have to be modified to accommodate peculiarities specific to particular foods. Compromises in sample preparation and the limitations of different instruments poses difficult questions, for example whether to exclude specular reflection and/or include the edge effects of translucent materials (Atkins and Billmeyer, 1966; Hunter, 1975; MacDougall, 1983). Although different spectrophotometers and colour meters are standardised identically this is no guarantee that they will produce the same colour values even if the variables associated with the foods are also standardised (Kent, 1987), for example both solids concentration and instrument geometry are important in grading tomato paste (Brimelow, 1987). In a study of evaporated milk, MacDougall (1987) showed that the ratio of absorption to scatter in the product is affected separately by dilution with water and the aperture area of the instrument. Hence, colour measurement of foods, especially if translucent, must be related to visual judgement. Reliance on physical data alone could lead to erroneous interpretation of the food's colour and appearance.

COLOUR SPECIFICATION

The factors required to specify colour are the spectrum of the illuminant, the reflection or transmission spectrum of the object and the colour matching functions of a standard observer. The calculation procedure is given in any standard text (Judd and Wyszecki, 1975; Wright, 1980; Hunt, 1987). The system is based on the trichromatic principle of the detection mechanism of the foveal cones in human vision but instead of using "real" primaries, red, green and blue, it uses X, Y and Z as "imaginary" primaries and a chromaticity diagram with coordinates $x = X/(X + Y + Z)$ and $y = Y/(X + Y + Z)$. The system was constructed in such a way that primary Y contains the entire lightness stimulus. With the 1931 CIE chromaticity diagram, it became the colour space to which all others are referred. Every colour can be uniquely located in 1931 CIE space by Y and x,y provided the observer and the illuminant are specified. The original 2° standard observer was supplemented in 1964 with 10° colour matching functions which are still trichromatic if related to tasks carried out in high intensity light. Until recently CIE illuminant C, representative of average daylight (6774°K), was the most used reference illuminant but D_{65} (6500°K), which includes some near ultra-violet, has replaced it. The current trend is to use D_{65} and 10° (ASTM,1987; CIE, 1987).

Although the CIE Y, x, y system specifies colour in three dimensions it is not uniformily spaced. Lines of constant hue distort to curves, constant chroma to ovals and equal visual distances increase several-fold from purple to green (MacAdam, 1942; Stiles, 1946). Transformations of Y, x, y have improved uniformity but it is unlikely that an ideal uniform space is attainable (Billmeyer and Saltzman, 1981). Three near-uniform colour spaces of practical importance are the Hunter (1958) L, a, b opponent colour space and the 1976 CIELUV and CIELAB spaces (Robertson, 1977). Studies on their reliability in comparison with newer scales used in the textile industry indicate that they have serious disadvantages for accurate colour matching (McDonald, 1985; McLaren, 1981).

Of the two 1976 CIE spaces CIELAB is the most generally applicable. The lightness coordinate (L*) is the same for both but no simple relationship exists between the chromaticness diagrams. CIELAB (L*, a*, b*) is a nonlinear cube root transformation of X,Y,Z to approximate the Munsell system whereas CIELUV (L*,u*,v*) has a linearly transformed chromaticity diagram similar to the 1931 space in which two component additive mixtures lie on straight lines. The formulae for CIELAB are:

$$L^* = 116(Y/Y_0)^{1/3} - 16$$

$$a^* = 500\{(X/X_0)^{1/3} - (Y/Y_0)^{1/3}\}$$

$$b^* = 200\{(Y/Y_0)^{1/3} - (Z/Z_0)^{1/3}\}.$$

where X_0, Y_0, Z_0 are the nominally white object colour stimulus. Psychometric hue (h*) and chroma (C*) are calculated by:

$$h^* = \tan^{-1} (b^*/a^*)$$

$$C^* = (a^{*2} + b^{*2})^{1/2}.$$

Total colour difference (ΔE^*) can be expressed either in the coordinates of colour space or in the correlates of lightness, chroma and hue.

$$\Delta E^* = \{(\Delta L^*)^2 + (\Delta a^*)^2 + (\Delta b^*)^2\}^{1/2}$$

$$\Delta E^* = \{(\Delta L^*)^2 + (\Delta C^*)^2 + (\Delta H^*)^2\}^{1/2}$$

where ΔH^* is used rather than Δh^* because the latter is angular. For small differences away from the L^* axis with h^* in degrees

$$\Delta H^* = C^* \Delta h^* (\pi/180).$$

OPAQUE AND TRANSLUCENT COLOURS

The reflectance of opaque or translucent foods depends on attenuation by pigment absorption, scatter by the internal structure, and directional scatter from the surface (MacDougall, in press). Colour calculated from the reflectivity spectrum, and the angular distribution of the surface's specular component are usually sufficient information to describe appearance if the object is flat and opaque, for example paint films where virtually all the reflectance is returned from the upper 1 mm. Although few foods are sufficiently flat for specular reflection to be measured accurately the components of glossiness or mattness must be accounted for in the measurement. Most colour meters are designed to eliminate specular reflection and most spectrophotometers have the option of including or excluding the specular.

Colour measurement by itself is inadequate to describe non-opaque materials, even if they are only slightly translucent. To specify translucency, values for absorption, scatter, and internal transmittance are required. The relative contributions that absorption and scatter have on reflectance can be separated by the two-variable procedure of Kubelka and Munk (Kubelka, 1948; Allen, 1978; Judd and Wyszecki, 1975). This relates reflectivity (R_∞), the reflectance at infinite thickness, to the absorption and scatter coefficients, K and S, by

$$K/S = (1 - R_\infty)^2 / 2R_\infty$$

where K and S are calculated from the reflectance of thin layers on white and black backgrounds. Correction factors to account for reflections at the air/object boundary (Saunderson, 1942; Allen, 1978) can be included but are probably unnecessary for foods. Examples of where the Kubelka-Munk analysis has been used on foods include pigment oxidation in fresh meat, translucence in bacon, and dilution of orange juice (MacDougall, 1982; 1983; in press).

TRANSPARENT COLOURS

It is easier to measure the colour of transparent beverages than foods that are opaque or translucent. However, conversion of transmission spectra into colour values does not match perceived colour because the pathlength viewed is usually much greater than that measured (MacDougall, 1987). Objective measurement of distilled beverage colour is not the usual practice adopted for classifying their colour. Procedures based on absorbance, using either narrow band filters or monochromatic light, have been recommended for alcoholic drinks which may include added caramel (Strunk et al., 1981).

In preparation for this symposium a selection of whiskies and caramel solutions were measured and their colours related to absorbance. The samples were selected as representative of those encountered by the trade (Table 1). They ranged from a clear "white" new distillate, through an immature malt, to two aged single malt whiskies, one somewhat paler than the other, to a typical blended Scotch whisky and a American Bourbon. Two samples of caramel, of similar concentration but different tinctorial power, and a sample of Paxarette, used to impart brownness to whisky, were also measured.

Table 1. Absorbance (log1/T) at 520nm and 10 mm for distilled beverages, caramel solutions and Paxarette and CIELAB lightness (L*) at 10 and 40mm.

	log1/T 10mm	L* 10mm	L* 40mm
Glen Grant:10 years old	0.07	95.3	83.9
Glen Farclas: 8 years old	0.15	89.8	68.6
Johnnie Walker: Black Label	0.17	89.5	67.8
Medley's Kentucky Bourbon	0.24	85.3	55.5
Braes of Glenlivit: new distillate	<0.01	99.2	98.8
Immature Malt	0.04	97.3	91.5
Paxarette: 2% in 40% alcohol	0.05	96.3	87.6
Caramel (A): 0.1% in 40% alcohol	0.19	88.2	67.4
Caramel (B): 0.1% in 40% alcohol	0.22	86.4	61.1

Percent transmittance from 380 to 770 nm was measured on a Pye Unicam SP 800 spectrophotometer at 40, 20, 10, 5, 2 and 1 mm pathlengths. The spectra were converted to CIELAB D_{65} (10°) uniform lightness (L*), chroma (C*) and hue angle (h*). Values of C* and h* for the pale malt and deeper coloured blended whisky were also calculated for tungsten filament light and three CIE fluorescent illuminants (CIE, 1987) using 20 nm weighting coefficients (ASTM, 1987). CIELAB C* and h* for these illuminants is corrected for visual adaptation by locating the illuminant at the achromatic axis of the space.

The relationship of absorbance (log 1/T) at 520 nm to lightness (L*) for 10 mm pathlength was found to be linear (Figure1). Chroma (C*) also

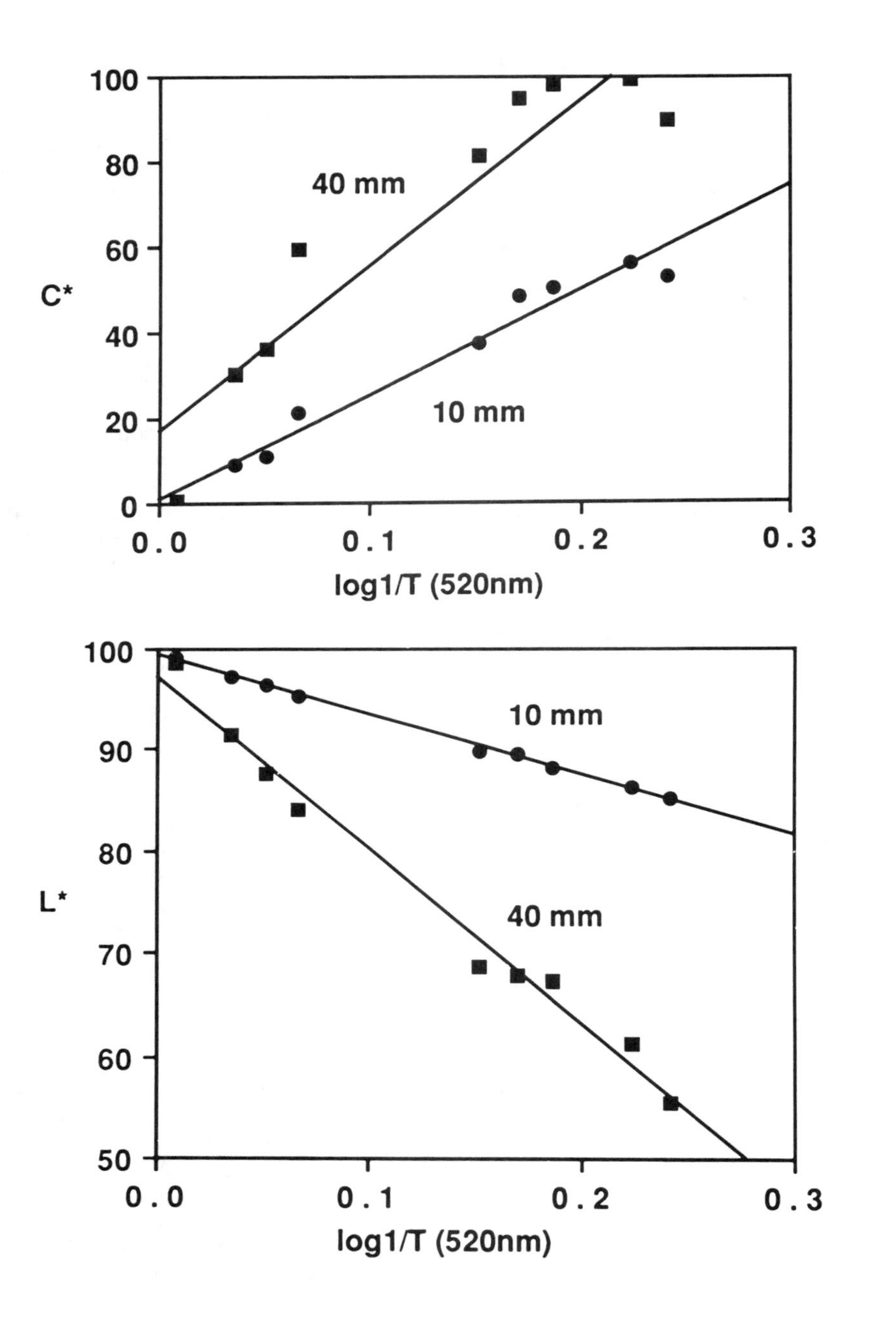

Figure 1 Relationship of absorbance (log 1/T) at 520 nm to CIELAB lightness (L*) and chroma (C*) at 10 and 40 mm pathlengths for the samples listed in Table 1.

increased with absorbance but with indication of possible decrease in C* for highly absorbing samples. Decrease in C* after its initial increase as L* increases is expected in dyed textiles (McLaren, 1986) and in highly pigmented beverages (MacDougall, 1987). Hence, single wavelength measurements might suffice as indicators of lightness but not of other colour characteristics. Which wavelengths to use has been discussed by Strunk et al. (1981) who conclude that those close to maximum absorption (approximately 430 nm) should be avoided. Small deviations in spectrophotometer calibration could incur large errors because caramel extinction rapidly changes in slope in the "blue" region of the spectrum. Also, correlation with lightness improves if absorbance is measured at wavelengths near maximum sensitivity of human vision. Typical transmittance values at 420 to 440 nm are 15 to 60 percent, equivalent to absorbance of 0.8 to 0.2, whereas at 520 to 540 nm transmittance values are 50 to >80 percent, equivalent to absorbance of 0.3 to <0.1.

The inter-relationships of CIELAB L*, C* and h* over all pathlengths for the two malt and the blended whiskies were compared to caramel (A) which was indistinguishable visually from the blended whisky. The spectra, from which L*, C* and h* were calculated, are shown in Figure 2 and the colour inter-relationships in Figure 3. As lightness decreased, with increase in pigmentation or pathlength, the hue angle decreased, indicating an increase in yellowness or brownness. Chroma increased with decrease in lightness indicating greater depth of colour. Over all pathlengths L* and C* of the four samples were related to h* as superimposed curves but the presence of caramel did affect the relationship of L* to C*. The caramel solution was virtually identical to the blended whisky but the malt whiskies had lower chroma relative to lightness, especially at pathlengths approaching those that would be viewed in a filled glass. The size and direction of the shifts in chromaticness, the combined effect of C* and h*, produced by the spectral quality of the light are similar for both the pale malt and the deeper colour of the blended (Figure 4). These values should be compared with the C* by h* diagram for D_{65} (Figure 3). There was little difference in chromaticness between D_{65}, cool white and natural fluorescent because whisky spectra have no sharp secondary absorption bands. Increasing red content in the lamp's spectral distribution to warm white and then to tungsten separately moved h* by 5°. Although there were only minor changes in chroma among the fluorescent lamps tungsten produced a 5 unit decrease in C* which, like the 5° decrease in h* redwards, is usually accepted as sufficiently large to remember.

The considerable importance of colour in whisky quality is confirmed in the paper in this symposium by Guy et al. (1988). Addition of "extra" caramel to a well known blended whisky influenced untrained consumer response by increasing the overall appreciation of the product.

ACKNOWLEDGEMENTS

The author wishes to thank Dr S. White of Strathclyde University for supplying the whisky samples and Miss S. Raine for her assistance in measuring them.

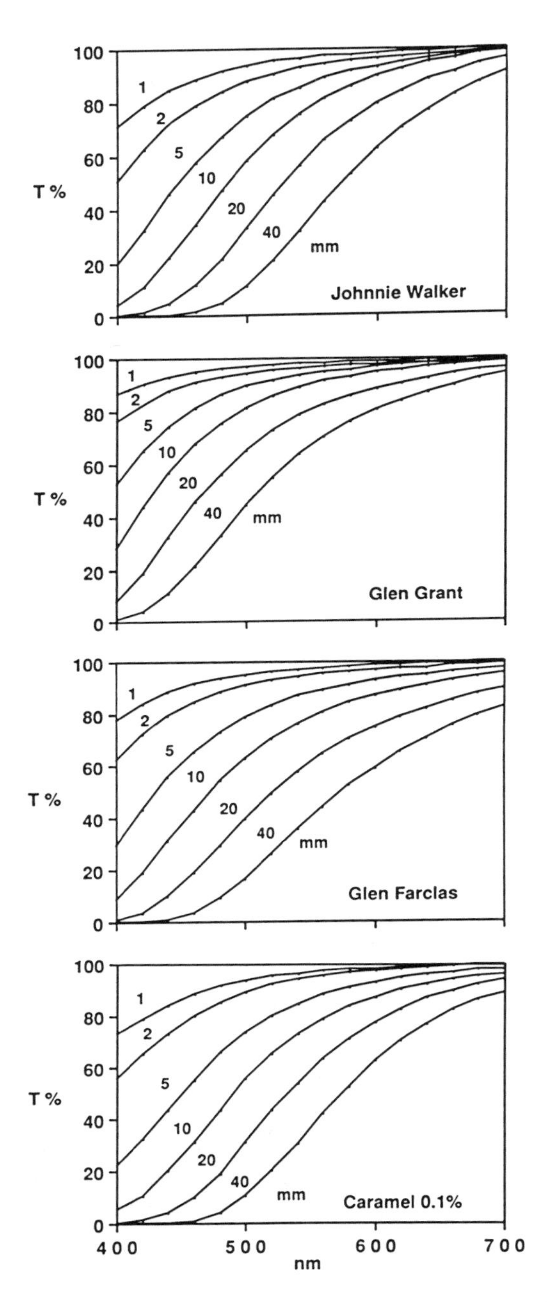

Figure 2. Transmission spectra of three whiskies and a similarly coloured caramel solution at pathlengths from 1 to 40 mm.

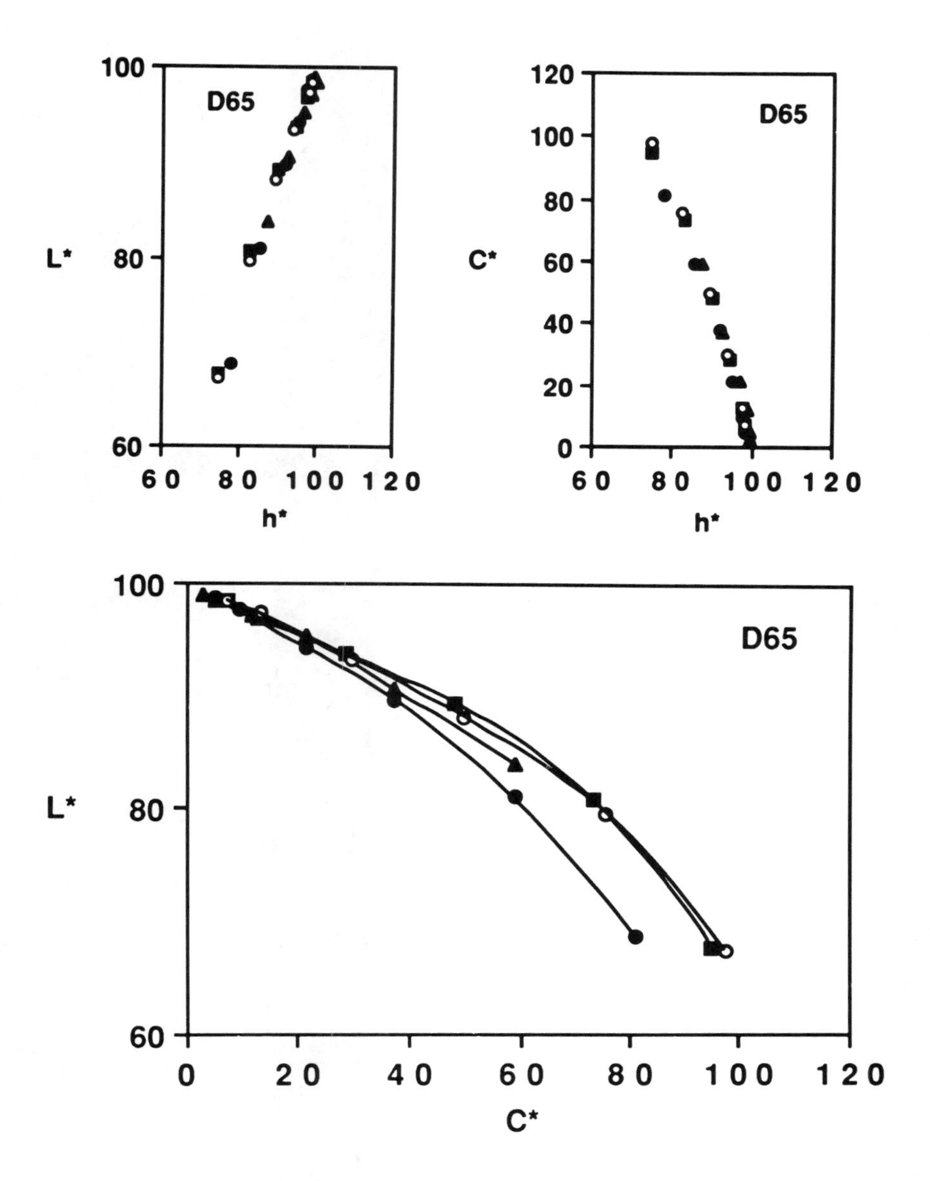

Figure 3. Inter-relationships of CIELAB lightness (L*), chroma (C*) and hue angle (h*) for three whiskies, Glen Grant (▲), Johnnie Walker (■) and Glen Farclas (●) and caramel (○) over pathlengths 1 to 40 mm.

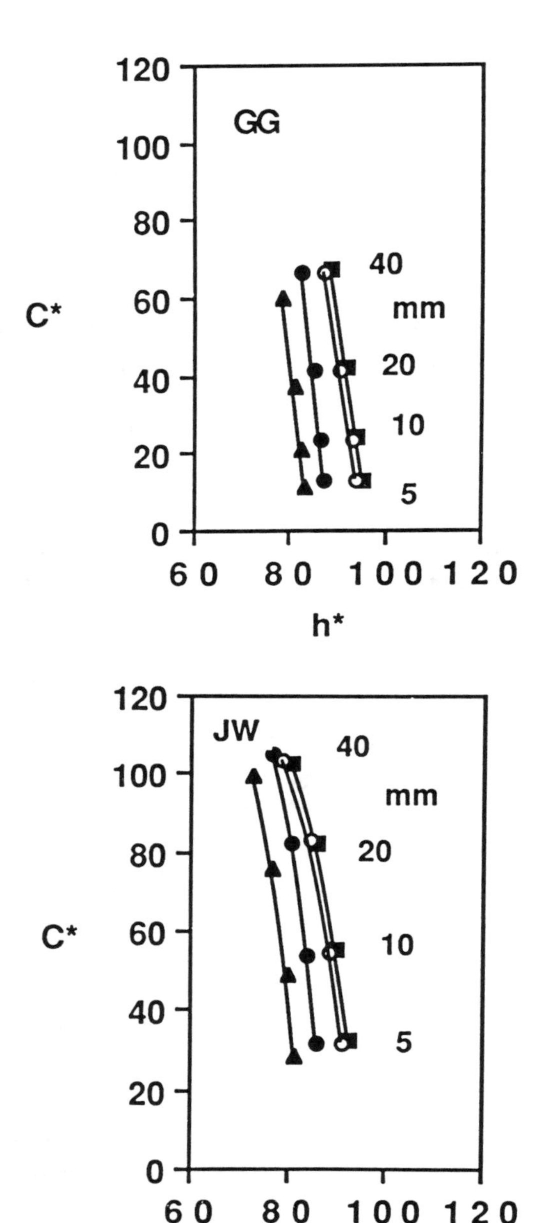

Figure 4. Calculated changes in CIELAB chroma (C*) and hue angle (h*) for pale coloured Glen Grant malt (GG) and darker blended Johnnie Walker (JW) from 5 to 40 mm pathlengths for four CIE illuminants, Source A Tungsten filament (▲), Fluorescent: Natural (■), Warm White (●) and Cool White (○).

REFERENCES

Allen, E. (1978). Advances in colorant formulation and shading. In: *AIC Color 77,* Adam Hilger, London, 153-79.

ASTM (1987). *Standards on Color and Appearance Measurement - E308,* (2nd edn), American Society for Testing and Materials, Philadelphia.

Atkins, J.T. and Billmeyer, F.W. (1966). Edge-loss errors in reflectance and transmittance measurement of translucent materials. *Mater. Res. Stand.*, **6**, 564-9.

Billmeyer, F. W. and Saltzman, M. (1981). *Principles of Color Technology*, (2nd edn), John Wiley and Sons, New York.

Brimelow, C. J. B. (1987). Measurement of tomato paste colour: Investigation of some method variables. In: *Physical Properties of Foods - 2: COST 90 bis Final Seminar Proceedings,* Elsevier Applied Publishers Ltd, London, 295-317.

CIE (1987). *Colorimetry*, (2nd edn), Commission Internationale de L'Éclairage, Vienna.

Guy. C., Piggott, J. R., Marie, S. and Conner, J. (1988). Consumer profiling of Scotch whisky. In: *Distilled Beverage Flavour: Recent Developments,*. Ellis Horwood Publishers, Chichester.

Hunt, R. W. G. (1978). Colour terminology. *Color Res. Appl.*, **3**, 79-87.

Hunt, R. W. G. (1987). *Measuring Colour,* Ellis Horwood Publishers, Chichester.

Hunter, R. S. (1958). Photoelectric color difference meter. *J. Opt. Soc. Am.,* **48**, 985-95.

Hunter, R. S. 1975). *The Measurement of Appearance,* John Wiley and Sons, New York.

Judd, D. B. and Wyszecki, G. (1975). *Color in Business, Science and Industry,* (3rd edn), John Wiley and Sons, New York.

Kent, M. (1987). Collaborative measurements on the colour of light scattering foodstuffs. In: *Physical Properties of Foods - 2: COST 90 bis Final Seminar Proceedings,* Elsevier Applied Publishers Ltd, London, 277-294.

Kubelka, P. (1948). New contributions to the optics of intensely light scattering materials. *J. Opt. Soc. Am.,* **38**, 448-57.

MacAdam, D. L. (1942). Visual sensitivities to color differences in daylight. *J. Opt. Soc. Am.,* **32**, 247.

MacDougall, D.B. (1982). Changes in the colour and opacity of meat. *Food Chemistry,* **9**, 75-88.

MacDougall, D. B. (1983). Instrumental assessment of the appearance of foods. In: *Sensory Quality in Foods and Beverages: Its Definition, Measurement and Control,* Ellis Horwood Ltd, Chichester.

MacDougall, D. B. (1985). The chemistry of colour and appearance. *Food Chemistry,* **21**, 283-299.

MacDougall, D. B. (1987). Effects of pigmentation, light scatter and illumination on food appearance and acceptance. In: *Food Acceptance and Nutrition,* Academic Press, London.

MacDougall, D. B. (1987). Optical measurements and visual assessment of translucent foods. In: *Physical Properties of Foods - 2: COST 90 bis Final Seminar Proceedings,* Elsevier Applied Science Publishers Ltd, London, 319-330.

MacDougall, D.B. (In press). Colour vision and appearance measurement. In: *Sensory Analysis of Foods,* (2nd edn), Elsevier Applied Science Publishers Ltd, London.

McDonald, R. (1985). The CMC colour difference formula and its performance in acceptability and perceptibility decisions. In: *AIC Mondial Couleur 85,* Monte Carlo, **1**, 46.

McLaren, K. (1981). The development of improved colour-difference equations by optimisation against acceptability data. In: *The Golden Jubilee of Colour in CIE,* The Society of Dyers and Colourists, Bradford.

McLaren, K. (1987). *The Colour Science of Dyes and Pigments,* (2nd edn), Adam Hilger, London, 157.

Robertson, A. R. (1977). The CIE 1976 colour-difference formulae. *Color Res. Appl.,* **2**, 7-11.

Saunderson, J. L. (1942). Calculation of the color of pigmented plastics. *J. Opt. Soc. Am.,* **32**, 727-36.

Stiles, W. S. (1946). A modified Helmholtz line element in brightness-colour space. *Proc. Phys. Soc. (London),* **58**, 41.

Wright, W. D. (1980). *The Measurement of Colour,* (5th edn), Adam Hilger Ltd, London.

9

Role of the continuous distillation process on the quality of Armagnac

Alain Bertrand
Institut d'Oenologie, Université de Bordeaux II,
351 cours de la Libération, F-33405 Talence, France

INTRODUCTION

Armagnac owes its quality to the "terroir" and to the different grape cultivars used, but its major character certainly results from the distillation process. The Armagnac still appeared two centuries ago [1] and 90% of Armagnac is produced with this process. However, double distillation (batch distillation), with the charentais pot still, was allowed again in 1972.

Continuous distillation is not expensive, in terms of energy, and functions more quickly than batch distillation, but it gives a very different spirit depending on (a) the volume of the boilers, (b) the number of plates in the distillation column, especially "dry plates" (plates which are above the point of entry of the wine), (c) the presence of a coil at the head of the column, (d) the use of a condenser for tailings, and (e) the use of a condenser for head products at the level of the wine heater. The distiller works empirically to obtain the required spirit.

Research to increase quality needs a good knowledge of the product in order to propose beneficial modifications. Traditional analyses do not give sufficiently detailed information for obtaining an idea of the intensity of the different aromas, and work dealing with volatile substances of wine spirits is rare [2-8]. We have therefore analysed for forty compounds by gas chromatography.

Initially, we decided to take a large number of samples of wines and the corresponding distillates in order to have as many differentiation criteria as possible. The samples were those of current practice. From this study it appears that Armagnac is particularly rich in certain compounds which distill only partially, due to their polarity or molecular weight. This is very clear when we compare the two processes of distillation, batch and continuous, for substances such as 2,3-butanediol, ethyl lactate and phenethyl alcohol. These compounds give Armagnac a vinous odour, a rough taste which balances a strong aroma coming from ripe grapes, but which needs long aging. For economic reasons it must be possible to commercialize the spirit as soon as possible; this requires better rectification allowing for a shorter aging period. The relatively high concentration of "tail" products is due to continuous distillation. The Armagnac still works on a "vapour flushing" principle. It is provided with different accessory systems (Figure 1) in order to better rectify the distillate.

The second part of our study deals with the composition of these tailings and the evolution of their composition in relation to the place where they are drawn from the still (condenser, coil, etc.) and the method of running the still.

One important parameter is the alcoholic strength of the spirit, which varies from 52 to 72% vol. The distillation of "tail" products obeys mathematical laws; their concentration varies exponentially with the alcoholic strength of the spirit.

Research now underway is attempting to define the true effect of the substances involved on the spirit and its organoleptic qualities.

ANALYTICAL METHODS

Volatiles of wines and spirits were analysed by gas chromatography, malolactic fermentation was controlled by H.P.L.C., and the percentage of lees of the must was measured by centrifugation (Alpha-Laval).

Higher alcohols, methanol, ethyl acetate, ethanol and acetal were determined by direct injection onto a classical column of Carbowax 400 Hallcomid (7.5 m x 2.2 mm) [9]. The internal standard was 4-methyl-2-pentanol.

Acetoin, ethyl lactate, 2,3-butanediol (D+Meso), and diethyl succinate were analysed with a capillary Carbowax 20M column (25 m x 0.32 mm). One ml of wine

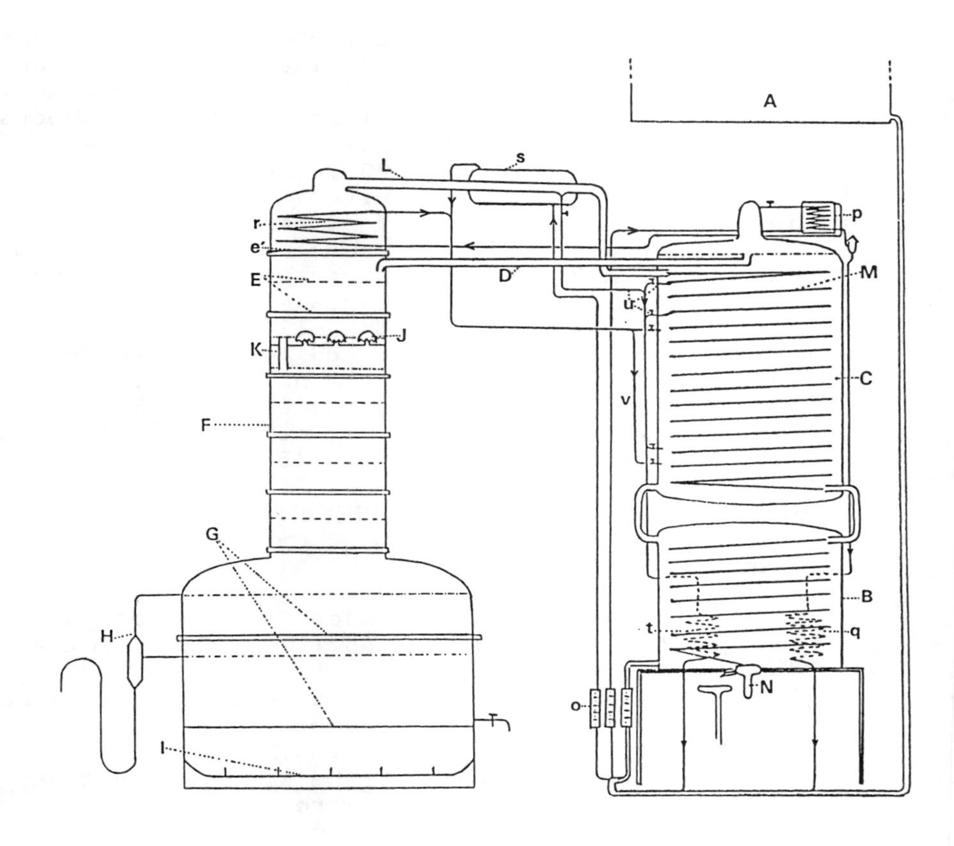

Figure 1: Alambic armagnacais (Armagnac still)

A cuve de charge (head of wine). B réfrigérant (cooler). C chauffe-vin (wine heater). D tubulure d'arrivée du vin dans la colonne (wine arrival). E plateaux (plates). F colonne (column). G chaudières (boilers). H siphon d'évacuation des vinasses (washy wine siphon). I foyer, rampe à gaz (furnasse). J calotte destinée au barbotage des vapeurs (half sphere). K tubulure de trop-plein du vin (overflow-pipe). L col de cygne (swan neck). M serpentin (coil). N porte alcoomètre (alcoholmeter holder).

Annexes (accessories): e plateau sec (dry plate). o débimètres du vin (wine debimeters). p condenseur de ttes (head condenser). q réfrigérant de ttes (head coil). r réfrigérant en tte de colonne (head column coil). s condenseur de produits de queue (tailings condenser). u tirage et recyclage de produits de queue (drawing and recycling of tailings). v recyclage du vin sortant des annexes (recycling of wines coming from accessories).

or spirit was added to 2 ml of methanol and 5 ml of ether; internal standards were 3-octanol and 1,4-butanediol (Figure 2).

Esters, fatty acids, phenethyl alcohol and hexanol were extracted from 50 ml of wine or spirit diluted to 1/5 by 4, 2 and 2 ml of a mixture of ether and hexane (1:1, v/v). Internal standards were 3-octanol and heptanoic acid. The analytical column was a capillary F.F.A.P. (50 m x 0.22 mm).

Diacetyl was determined after dilution of 1 ml of wine or spirit with 2 ml of methanol and 5 ml of ether. A classical celanese column (6 m x 2.2 mm) was used with an electron capture detector. The internal standard was 2,3-pentanedione.

Aromatic aldehydes (a), phenolic acids, hydroxymethyl furfural, furfural (b) and scopoletine (c) were simultaneously determined by H.P.L.C. with a C18 bounded phase (5 µm) column (18 cm x 5 mm). The detectors were respectively UV at 313 nm (a), UV at 280 nm (b) and fluorimeter at 545 mn (c).

ANALYSIS OF 58 WINES AND 80 SPIRITS

Experimental

We tried to differentiate wines in relation to

(a) cultivars (Ugni blanc, Baco 22 A, Folle Blanche, Colombard),

(b) absence or presence of settling of the must,

(c) addition of dried yeasts,

(d) malolactic fermentation,

(e) time of resting of the wine on the lees.

For the spirits, criteria were

(a) content of the boilers,

(b) number of plates,

(c) alcoholic strength of the spirit,

(d) possibility of drawing head and tail products.

Some wines were distilled with 2 or 3 different stills. Analytical results were statistically processed (P.C.A.) by PAGES (E.N.S.A., Rennes).

Results

On comparison of averages between wines and spirits (Table 1), as a first approximation we can say that

(a) higher alcohols distil completely with the exception of phenethyl alcohol,

(b) esters distil as completely with the exception of diethyl succinate and ethyl lactate. Heat causes release of ethyl octanoate and ethyl decanoate from yeast and this explains the exceptional recovery (above 100%),

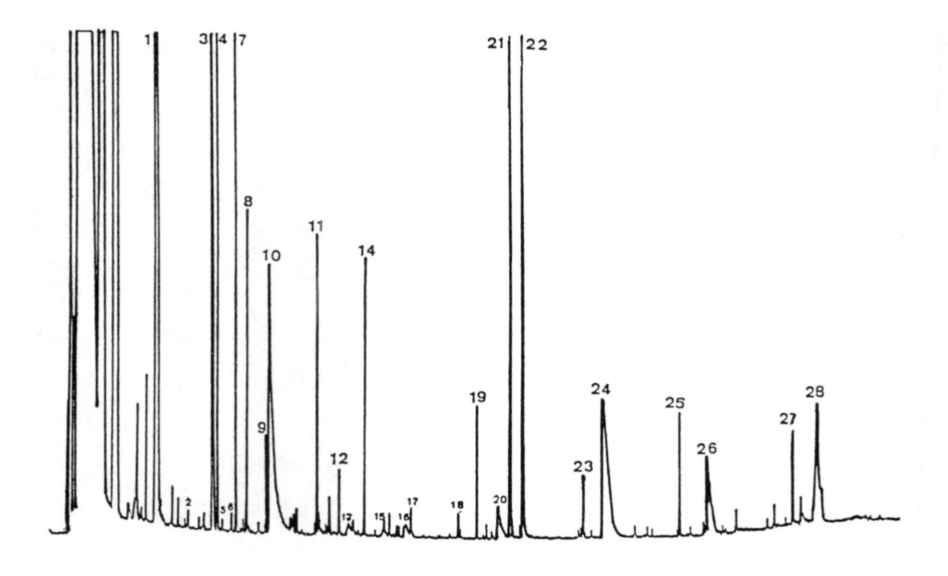

Figure 2: Direct injection of Armagnac in C20M capillary column 1 3-Methyl 1-butanol. 2 Acetoin. 3 Ethyl lactate. 4 1-Hexanol. 5 t.3-hexene-1-ol. 6 c.3-hexene-1-ol. 7 3-Octanol (internal standard). 8 Ethyl Octanoate. 9 Furfural. 10 Acetic acid. 11 2,3-Butanediol (D-). 12 2,3-Butanediol (meso). 13 Isobutyric acid. 14 Ethyl decanoate. 15 Butyric acid. 16 Diethyl succinate. 17 Isovaleric acid. 18 Phenethyl acetate. 19 Ethyl dodecanoate. 20 Hexanoic acid. 21 Phenethyl alcohol. 22 1,4-(internal standard). 23 Ethyl tetradecanoate. 24 Octanoic acid. 25 Ethyl hexadecanoate. 26 Decanoic acid. 27 Ethyl octadecanoate. 28 Dodecanoic acid.

Table 1: Comparison of averages in wines and spirits

	Wines	Spirits	% recovery
Ethanol (% vol.)	11.1	59.7	-
Methanol	41.4	194	87
1-Propanol	30.7	161	98
2-Methyl 1-propanol	66.6	377	105
2-Methyl 1-butanol	51.6	285	103
3-Methyl 1-butanol	224	1220	101
1-Hexanol	1.75	10.5	111
2-Phenyl ethanol	52.9	31.4	11
Ethyl acetate	40.9	207	94
Isoamyl acetate	2.25	10.9	90
Hexyl acetate	0.11	0.65	10
2-Phenyl acetate	0.26	1.18	84
Ethyl hexanoate	0.46	2.57	104
Ethyl octanoate	0.69	5.01	203
Ethyl decanoate	0.31	6.43	385
Diethyl succinate	1.62	3.77	43
Ethyl lactate	340	248	14
Isobutyric acid	1.34	3.2	44
Isovaleric acid	0.69	0.92	25
Hexanoic acid	3.08	4.08	25
Octanoic acid	4.96	20.02	75
Decanoic acid	1.13	6.01	99
Diacetyl	0.65	2.91	83
Acetoin	7.22	30.2	78
2,3-Butanediol	549	14.6	0.5

Results in mg/l except indication.

(c) volatile acids distil at a level of 25 to 30% with the exception of butyric acid (this can be a systematic error in the measurement), if it is accepted that octanoic and decanoic acids are released by yeast in the same proportions as the corresponding esters,

(d) diacetyl and acetoin certainly distil completely but determinations were not precise,

(e) 2,3-butanediol distils at a level of only 0.5%. However this is very high, and Armagnac, with a content of 14 mg/l is certainly the spirit which is the richest of all in butanediol. Lafon _et al._ [2]

maintained the optimal concentration of 2,3-butanediol with regard to the quality of Cognac was about 1.0 to 1.5 mg/l.

Multidimensional analyses

Principal components analysis (P.C.A.) [10] did not allow for differentiation of the cultivars. This is normal if one considers that the substances which are determined come from yeast or bacteria. The settling of the must increases esters and diminishes higher alcohols.

Malolactic fermentation could be correlated with ethyl lactate.

P.C.A. results for some wines and corresponding spirits are shown together with their variables (Figure 3). The first axis shows variables which are influenced by the vinification process; on the one hand, higher alcohols and on the other, fatty acids and their ethyl esters.

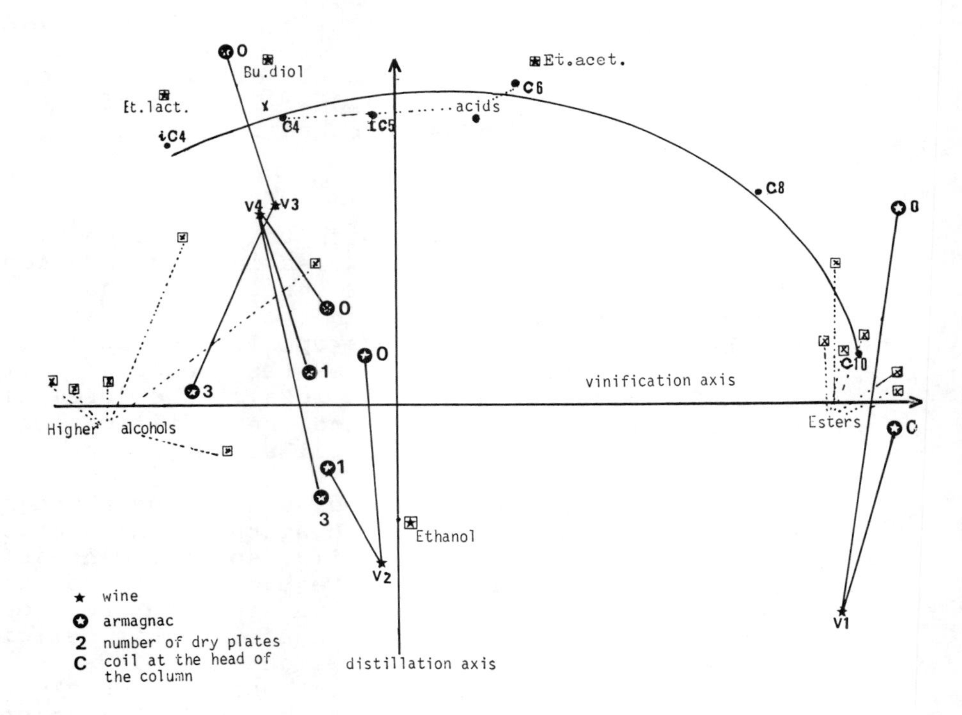

Figure 3: Principal components analysis of wines and spirits

The second axis shows ethanol versus head (ethyl acetate) and tail products (2,3-butanediol...). This axis is clearly the distillation axis.

When a wine is distilled with different stills, the corresponding spirits appear along lines which are parallel to the distillation axis, but never to the vinification axis. This is logical, and clearly demonstrates that vinification is decisive for the composition of the spirit. In any case, the aroma is more or less hidden by the content of the tail products. Of course the number of plates in the column and especially the number of "dry plates", or the use of a condenser, play a major role in the tail product content.

COMPARISON OF ARMAGNACS OBTAINED BY CONTINUOUS OR BATCH DISTILLATION

Experimental

We compared spirits obtained from the same wine distilled by

(a) a continuous still with 3 boilers i.e. with 2 plates in the boiler (SIER type),

(b) a continuous still with 2 boilers (SAVARY type), and

(c) a charentais still: (i) distillation of the "brouillis" alone, and (ii) distillation of "brouillis" + "secondes de bonne chauffe".

Results

Alcohols (Table 2). With the exception of phenethyl alcohol (which doubles in concentration in the continuous process), all the other higher alcohols distil completely in the "brouillis" on the one hand, and in the various spirits on the other.

Esters. Ethyl lactate is less abundant in the batch distillates (about 25% less). The "brouillis" is very rich in this ester and the rectification occurs during the second distillation (bonne chauffe).

Acids. The different acids are less abundant in batch distillation. Only one-sixth of the acetic acid of the "brouillis" distils, so sour wine is well rectified by this process; the same phenomenon is observed with ethyl lactate. Adding the "secondes" to the "brouillis" partially restores this deficit, at least for high molecular weight compounds.

Aldehydes and ketones. Ethanal, which gives a pungent odour, is greater with batch distillation. It seems that with continuous distillation a certain leak occurs, either because one part of the head is

Table 2: Comparison of Armagnacs distilled by batch or continuous distillation

	Continuous distillation		Batch distillation (double chauffe)		
	3 boilers	2 boilers	a	b	c
Ethanol (% vol.)	58.8	54.5	69.7	70.3	29.9
Total acidity (as acetic)	40.9	70.5	12.1	10.2	68.2
Methanol	43.3	39.2	40.8	38.9	40.7
1-Propanol	26.9	28.7	28.5	27.7	28.5
2-Methyl-1-propanol	88.4	92.0	103	90.7	97.8
2-Methyl-1-butanol	66.9	65.9	72.9	64.2	71.3
3-Methyl-1-butanol	242	253	286	266	284
1-Hexanol	1.55	1.74	1.81	1.52	1.59
c.3-hexene-1-ol	0.25	0.19	0.06	-	-
2-Phenyl-ethanol	6.94	7.42	2.98	3.52	25.53
Ethyl acetate	24.6	22.8	33.1	25.4	31.1
Isoamyl acetate	0.75	0.79	0.36	0.71	0.74
Hexyl acetate	0.08	0.08	0.07	0.08	0.19
2-Phenyl acetate	0.10	0.20	0.17	0.27	0.84
Ethyl butyrate	0.22	0.40	0.10	0.10	0.06
Ethyl hexanoate	0.29	0.28	0.34	0.26	0.27
Ethyl octanoate	0.54	0.56	0.73	0.59	0.47
Ethyl decanoate	0.40	0.39	0.56	0.44	0.28
Diethyl succinate	0.46	0.22	0.45	0.76	1.10
Ethyl lactate	50.5	39.0	34.0	34.7	108
Isobutyric acid	0.50	0.88	0.48	0.52	0.27
Isovaleric acid	0.32	0.26	0.17	0.27	0.84
Hexanoic acid	0.52	0.69	0.22	0.47	1.45
Octanoic acid	2.24	1.56	1.08	1.71	2.89
Decanoic acid	0.77	0.69	0.22	0.47	1.45
Diacetyl	0.68	0.80	0.91	0.80	
Acetoin	0.81	0.65	0.18	0.13	
2,3-Butanediol D-	2.34	3.11	0.07	0.06	6.85
2,3-butanediol meso	0.55	0.66	0.02	0.02	2.10
Ethanal	5.94	6.94	24.7	19.1	27.5
Acetal	5.41	3.32	8.10	5.82	4.53
Furfural	0.24	0.18	1.02	1.37	1.33

Results in g/hl pure alcohol except indication
a: bonne chauffe of the brouillis alone. b: bonne chauffe of the brouillis + secondes. c: brouillis.

not condensed and recycled or because this substance remains at the top of the wine heater. However, with batch distillation, it could be useful to have the distillate at a higher temperature when it runs out of the coil, in order to eliminate more of this substance. Furfural is more abundant in batch distillation spirits.

Acetoin is well rectified by batch distillation but diacetyl is the same with the two processes. 2,3-Butanediol is 30 times greater with continuous distillation; it is a real indicator of the distillation process.

TAILING PRODUCTS

In order to examine the effects of the different accessories of the stills we performed the following study.

Experimental

Tailings were taken from 7 different Armagnac stills from different parts of the still and in different conditions, representing a total of 19 samples. The samples were an aliquot of the tailings corresponding, at least, to the distillation of 400 l of spirit. All the distillation tailings were collected separately.

Tail drawings were made at different levels,

(a) "col de cygne" (the "swan neck"),

(b) condenser, and

(c) first or third turn of the "refrigerant" coil.

The alcohol content of the different spirits ranged from 55 to 64% vol. and the ratio between the flow of the spirit and the flow of the wine was from 0.15 to 0.17.

Results

After chromatographic analysis of the tailings which were summarized for the same group of substances we performed a P.C.A. of the results (Figure 4).

The first axis (33% of total variance) is defined positively by ethanol and substances which are highly correlated with ethanol such as higher alcohols, esters of fatty acids and fatty acids themselves, but with less importance for the latter; then, negatively by 2,3-butanediol, phenethyl alcohol and phenethyl acetate which are tail products.

The second axis represents secondary products of malolactic fermentation, particularly the nauseous acids, isobutyric and isovaleric, together with ethyl

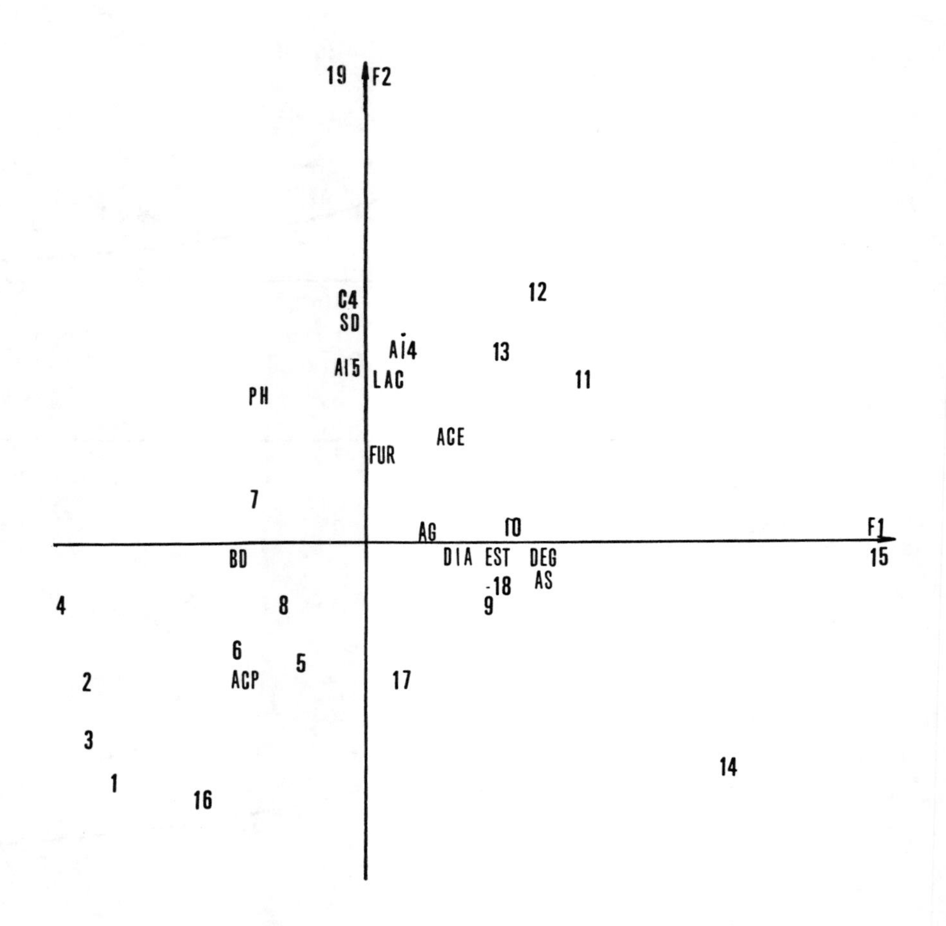

Figure 4: Principal components analysis of tail products 1-19 tailings. ACP phenethyl acetate. ACE acetoin. AG fatty acids (C6+C8+C10+C12). Ai4 isobutyric acid. Ai5 isovaleric acid. AS higher alcohols. BD butanediol (D- and meso). C4 butyric acid. DEG degree. EST esters (isoamyl acetate + fatty ethyl esters). FUR furfural. LAC ethyl lactate. PHE phenethyl alcohol. F1 : Axis 1. F2 : Axis 2.

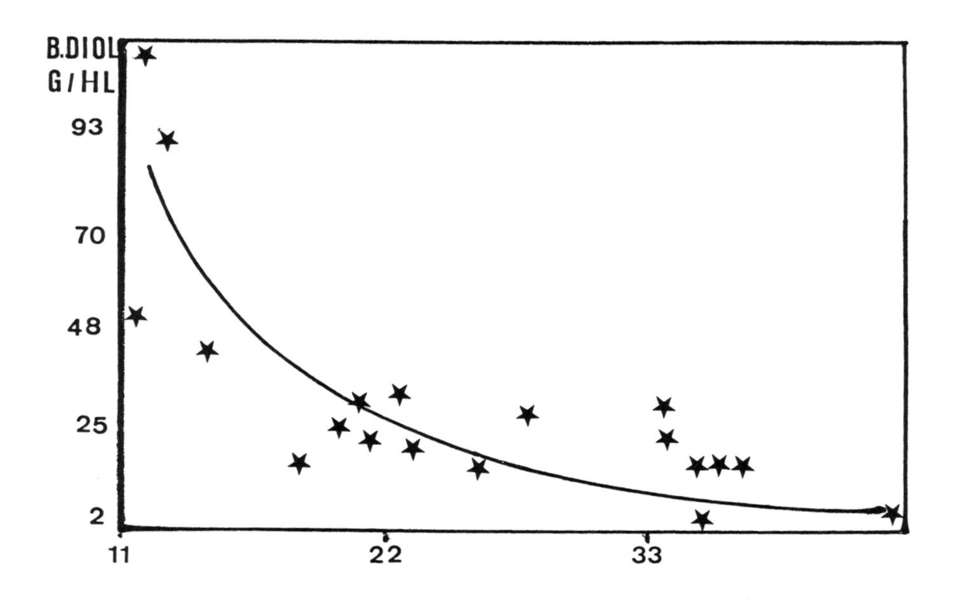

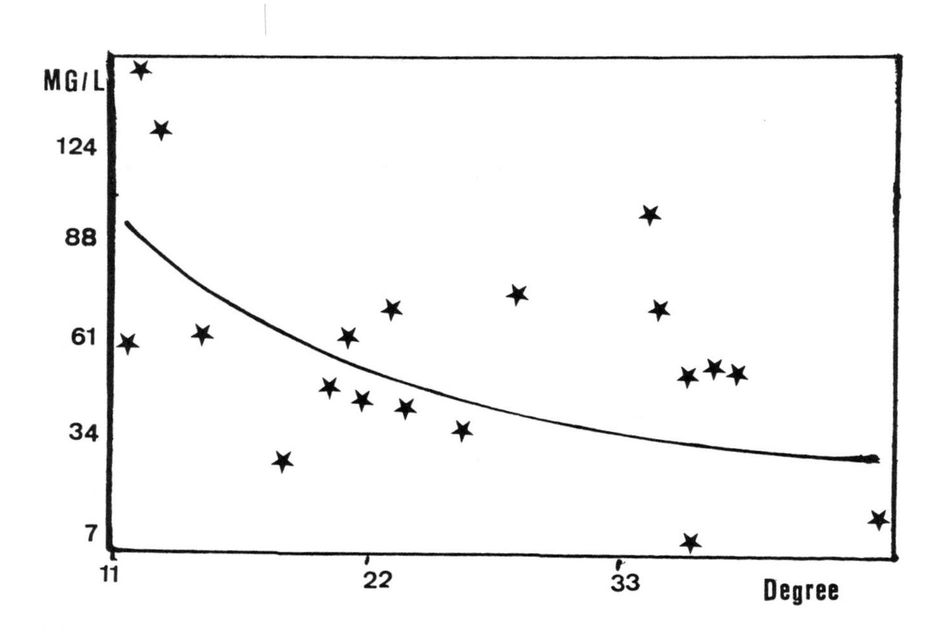

Figure 5: Evolution of butanediol in the tailings as function of the alcoholic strength of the tailings

lactate and diethyl succinate. Acetoin and phenethyl alcohol also play a positive role in this axis. It may be said that the first axis depends on distillation, while the second represents bacterial accidents (sample 19).

Samples 1, 2, 3, 4, are rich in tail products, about 75 g/hl pure alcohol (P.A.) of butanediol for an alcohol content of 13% vol. On the other hand distillates 14 and 15 have only 5 g/hl P.A. of butanediol for an alcohol content of 35% vol. The relationship between 2,3-butanediol and ethanol has the form

$$\text{Butanediol} = A \times (\text{alcohol content})^{B}$$

where A and B are constants.

This law is particularly clear when concentrations are expressed as g/hl P.A., which is not quite normal. However a similar law could be found when the results are expressed in mg/l (Figure 5) with a much greater scattering of the points.

Similar relationships have been found with other substances such as phenethyl alcohol. It could be said that removal (and subsequent recycling) of the tailings at a strength of 13% vol. makes it possible to reduce the butanediol content of the distillate by 4.7 mg/l (23% less). For samples 6, 13, 16 and 17, whose average strength is 22% vol., this decrease at the same level of withdrawal of the tailings is only 1.4 mg/l; the percentage reduction of butanediol then falls to 17.5%. For samples 14 and 15, whose alcohol content is 35% vol., butanediol is only diminished by 0.4 mg/l.

It must be noted that, for this calculation, the real alcohol content of the distillate (of the heart) is taken into account; thus samples 14 and 15, which are not purified by recycling of tailings, are nevertheless the lowest in butanediol and other polar compounds. This results from the use of a coil in the head of the column which is efficient enough to avoid additional recycling of tailings.

INFLUENCE OF VARIATIONS OF ALCOHOLIC STRENGTH ON THE COMPOSITION OF SPIRIT

Experimental

During distillation it is possible to modify the alcoholic strength of the spirit in two ways:

(a) by increasing the flow of the wine. Increasing the alcoholic strength, the calorific energy for the same volume of wine decreases. Only the most volatile substances are distilled (experiment A).

(b) by increasing the firing which causes a lowering of the alcoholic strength. Much calorific energy is used for the same volume of wine (constant flow rate). This has, as a consequence, the distillation of less volatile substances (experiment B).

Lastly, in experiment C, the flow of the wine and the temperature of heating were varied simultaneously. When the flow of the wine is increased in the head condenser (and consequently the total flow) this has the consequence of increasing the alcoholic strength too much; to correct this effect the firing is slightly raised.

Ethanol is the only compound which is easily controlled during distillation. Thus we compared the variation of the different substances to the alcoholic strength.

It is already known that higher alcohols and esters distil completely. We were therefore particularly interested in the fate of the tailings.

Results

Results are shown in Table 3. The decrease of the compounds in relation to ethanol are represented by curves which have the same shape (Figure 6). All these compounds which differ from each other in their chemical nature, nevertheless behave similarly with respect to ethanol; they diminish according to equations of the same type,

$$\text{substance} = A \times (\text{alcohol content})^{-5.9},$$

where A is a constant and -5.9 the average of the exponents.

All these substances are highly correlated with each other. For a given still, determination of one compound makes it possible to know with satisfactory precision the concentration of the others as well as their variation in relation to the alcoholic strength.

The non-linearity of the distillation of the tail products with regard to ethanol can explain significant variations in concentrations for small differences in alcoholic strength. A spirit distilled at 60% vol. can contain as much as twice the concentration of certain substances as one distilled at 65% vol. It may be understood that two spirits produced with the same still may differ greatly according to distillation conditions.

Table 3: Variation of tail products as function of the alcoholic strength of Armagnac

	Experiment A (flow of the wine)					Experiment B (heating)				Experiment C (flow + heating)			
Alcohol (% vol.)	52.1	56.4	59.9	66.1	70.7	54.9	61.2	65.7	68.0	57.5	62.3	67.3	72.3
2,3-Butanediol D	3.51	1.95	1.29	0.75	0.58	7.39	3.69	2.03	2.18	0.80	0.48	0.38	0.22
2,3-Butanediol M	0.99	0.54	0.37	0.23	0.19	2.03	1.09	0.81	0.60	0.20	0.07	0.04	0.02
Diethyl succinate	1.41	0.95	0.68	0.30	0.28	6.90	4.02	2.54	1.58	1.95	1.45	0.81	0.55
Furfural	0.70	0.28	0.06	0.03	0.01	2.54	2.24	0.59	0.58	0.70	0.28	0.03	0.01
2-Phenyl ethanol	7.41	4.76	3.38	1.95	1.33	35.0	21.4	16.9	13.6	17.3	11.2	6.4	4.20
Ethyl lactate	122	82.8	55.1	34.6	23.1	377	208	154	126	58	34	20	15
Acetic acid	-	-	-	-	-	-	-	-	-	2.81	2.09	1.20	1.16

Results in mg/l except indication.

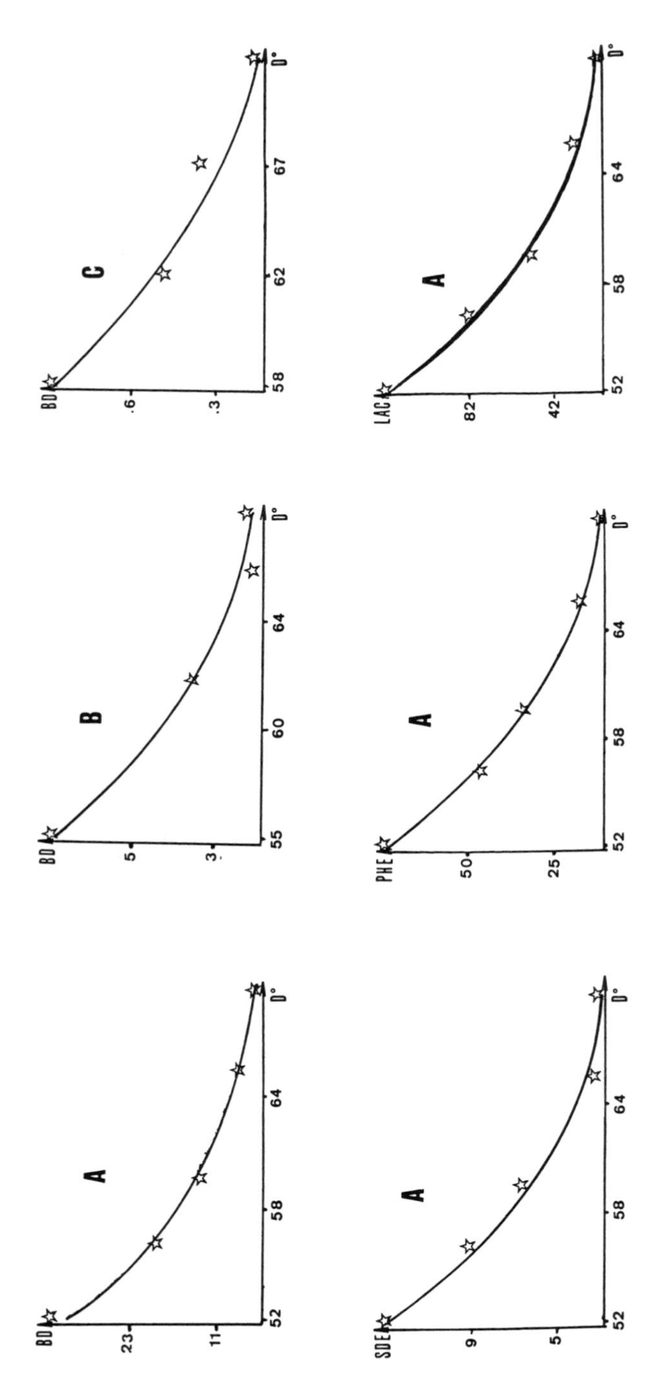

Figure 6: Evolution of tail products in Armagnac as function of the alcoholic strength. A, B, C experiments. BD Butanediol (D-). SDE diethyl succinate. PHE phenethyl alcohol. LAC ethyl lactate.

CONCLUSION

During continuous distillation with the Armagnac process the substances of wine behave differently, depending upon whether they are:

(a) very volatile, i.e. esters, higher alcohols. In this case they distil completely and depend only on the vinification process.

(b) polar compounds, e.g. 2,3-butanediol, phenethyl alcohol. In this case they diminish in relation to the strength of the spirit in an exponential manner. They also can be partially eliminated by withdrawal (and recycling) of tailings. However, to be efficient this process implies that the tailings have a low alcohol content. This results in a shorter aging period.

These observations could lead to better running of stills with a view to obtaining spirits with the exact composition required.

AN APPROACH TO THE ANALYTICAL TASTING OF ARMAGNAC

Colour

A direct light of medium intensity makes it possible to appreciate the colour, the shade of which goes from yellow green to golden yellow and red orange. Age plays a role as well as the wood but a red shade can often be correlated with hydroxymethylfurfural, which is present from 0.2 to 10 mg/l. This substance is more or less proportional to the caramel added to the spirit.

Odour

It is possible to note the different impressions as they appear.

Pungent. This odour can be correlated with the sum of ethanal and acetal, which varies from 50 to 200 mg/l. Gently shaking the glass reveals a certain burning of the nose from about 100 mg/l.

"Piqué". Ethyl acetate (100-500 mg/l) has an odour of glue solvent. The level of perception of this substance in Armagnac is about 250 mg/l. It is easier to recognise this odour when shaking the glass vigorously. It increases with age, and this is why the content in the young spirit must be low.

Fruity. This impression comes mainly from acetates of higher alcohols (especially isoamyl acetate), their concentration varying from 0.3 to 10 mg/l. These compounds are hydrolyzed during aging and disappear almost completely after 10 years.

Floral. Fatty acid ethyl esters (10 to 30 mg/l)

are mainly responsible for this odour.

Vanilla. This impression is caused by wood. Aromatic aldehydes contribute on a large scale to the aroma of the spirit. Vanillin is the most odorous. It varies from 0.2 to 7 mg/l in commercial Armagnac; it is clearly perceived at a level of 2 mg/l.

Grassy. This is reminiscent of cut grass. It is characterized by a high content of hexanol and 3-hexenol(s). This off-flavour is relatively rare in Armagnac, certainly because the grapes have a good maturity when they are harvested. The average is about 7 mg/l and the level of perception is about 12 mg/l.

Buttery. This is due to diacetyl at a level above 4 mg/l. Please note 2,3-butanediol varies in Armagnac from 3 to 20 mg/l with an average of 8 mg/l. At a high level this compound (and other tail products) can mask certain aromas: floral, pungent, woody, for example.

Taste

Sweet. The first impression to notice is the sugar. This varies from 0 to 8 g/l. It is tasted at the tip of the tongue. At a low level it delays the impression of alcohol burn, and at more than 4 g/l it can be tasted as itself.

Burning, dry. Some spirits present this off taste, which is difficult to quantify. Different factors can play a role: alcoholic strength, aldehydes, but also other volatile substances which are not yet known.

Astringency. It is caused by tannins of oak wood, tasted at the posterior part of the tongue. This can be well correlated with scopoletine, the level of which varies from 10 to 220 mg/l. Under 50 mg/l of scopoletine, astringency is generally not perceptible.

Intense aromatic persistance called P.A.I. in French. After having expectorated the spirit, and taking great care of the evolution of what remains in the mouth, we can count seconds before there appears a change in the sensations. A sort of dryness, a burning, a physiological modification at the level of the taste buds. These seconds represent the true length of the spirit. This is directly proportional to the aromatic richness of the spirit, and finally to its quality, if no off-flavour has already been detected.

Conclusion

These perceptions are the easiest to explain; however

there exists a range of odours and impressions which make it possible to better describe spirits, and which are generally found by tasters. For instance dry prune, musty, fusty, rancid, bitterness, etc.; however the chemical compounds responsible are not yet known or are very difficult to determine.

* This work has been financially supported by B.N.I.A. (Bureau National Interprofessionnel de l'Armagnac).

ACKNOWLEDGEMENTS

We would like to thank all those who contributed to this work: C. SANZ and M.C. SEGUR (B.N.I.A.), J. PAGES (E.N.S.A., Rennes), P. WILDBOLZ and Ph. JADEAU (Institut d'OEnologie).

REFERENCES

1. Dufor, H. Armagnac eaux-de-vie et terroir. (Privat Ed.), 1982.
2. Lafon, J., Couillaud, P., Gay-Bellile F. Le Cognac sa distillation. (Baillière Ed.), 1973.
3. Tourlière, S. Commentaires sur la présence d'un certain nombre de composés accompagnant l'alcool dans les distillats spiritueux rectifiés et eaux-de-vie. Ind. alim. agric. 1977.94, 565-574.
4. Caumeil, M. Le Cognac. Revue "Pour la Science", Scientific Americain éd. 1983, 48-57.
5. Wildbolz, P. Etude de critères analytiques en vue de l'amélioration des eaux-de-vie d'Armagnac. Mémoire D.E.A. 1986, Université de Bordeaux II.
6. Soufleros, E., Bertrand, A. Etude sur le "Tsipouro", eaux-de-vie de marc traditionnelle de Grèce, précurseur de l'Ouzo. Conn. Vigne Vin. 1987, 2, 93-111.
7. Jadeau, Ph. Incidence du débourbage des moûts et de la fermentation malolactique des vins sur la composition des eaux-de-vie d'Armagnac. Essai de caractérisation d'alambic continu. Mémoire D.E.A. 1987 Université de Bordeaux II.
8. Bertrand, A., Ségur, M.C., Jadeau, Ph. Analyse de la composition des queues d'eaux-de-vie d'Armagnac. Conn. Vigne Vin. 1988, 3, 89-92.
9. Bertrand, A., Boidron, J.N. Une nouvelle phase stationnaire pour l'étude des spiritueux par chromatographie en phase gazeuse. Conn. Vigne Vin. 1967, 2, 51-54.
10. Bertrand, A., Pagès, J., Analyse d'un échantillon de vins et d'eaux-de-vie d'Armagnac. Rapport B.N.I.A. Addad. 1986.

10

A scientific approach to quality control for Cognac spirits

Roger Cantagrel
Bureau National Interprofessionnel du Cognac,
F-16101 Cognac, France

SUMMARY

The quality of a spirit such as cognac results from factors incorporated at every step in the production process : from the diverse soils within the "appellation" (designated origin) area, the grape varieties, the cultivation of the soil and care of the vineyards, the vintage, the vinification, the storage of the wines before distillation, the distillation itself, the ageing, up to the bottling of the cognac before it is finally marketed.

Quality control may be performed at different stages in this process :

1 - On the must and the wine : measuring their level of possible microbial contamination (checking for laccase activity caused by Botrytis cinerea).

2 - On the wine : checking for possible defects by means of small-scale distillation in a laboratory. The major faults which may be revealed are : acetic, pricked, acrolein, putrid, oxidation, maderisation.

3 - <u>On newly-distilled spirits</u> :

- Better understanding of distillation and the order in which the various constituents pass over provides supplementary information for more effective control of the overall process.

- Analysis by capillary gas chromatography of some of the volatile compounds provides tasters with a tool to support their judgements. For compounds causing defects, prior determination must be made of flavour thresholds. Concentrations above these thresholds would indicate impairment of quality.

INTRODUCTION

Quality is incorporated into cognac at every step in the production process. The region of production and the growth localities into which it is divided (Grande Champagne, Petite Champagne, Borderies, Fins Bois, Bons Bois, Bois Ordinaires) are demarcated by French Government decree.

The region for growing the wines used to make cognac mainly covers two "departements" -Charente and Charente Maritime- along with two small zones within two other "departements" -Deux Sèvres and Dordogne. This region totals some 81000 hectares and consists mainly of white grape varieties, with Ugni-blanc, Folle Blanche and Colombard accounting for 90 % of the grape varieties used and Blanc Ramé, Jurançon Blanc, Montils, Semillon and select being used for the remaining 10 %.

Cognac has been granted an "Appellation d'Origine Controlée" (area of designated origin) and is thus subject to the strict production regulations laid down by the Institut National des Appellations d'Origine (INAO).

Cognac distillers have placed the Technical Department of the Bureau National Interprofessionnel du Cognac (Station Viticole) in charge of conducting research in any domain which might affect quality. Increasingly, the quality of cognac, one of the most prestigious of all spirits, must be unimpeachable.

This paper describes several methods which have been implemented to maintain cognac's high standard of quality.

I - DETERMINING LEVELS OF MICROBIAL CONTAMINATION IN WINES

High-quality spirits can only be achieved with grapes showing low levels of contamination. Current research aims at :

- ascertaining the qualitative affect on the spirits of grey rot in the vintage ;

- determining the contamination threshold beyond which the quality of the spirits is adversely affected.

We also wanted to evaluate whether a kit now being marketed for testing laccase activity was of practical value and survey the contamination level of the 1987 vintage by analyses on samples of wines submitted to us for small-scale distillation.

The fundamental method was the following : the must or wine is filtered through a syringe containing polyvinyl polypyrrolidone (PVPP) to fix the phenolic compounds. After buffering, the filtrate is reacted with syringaldazine. In the presence of laccase this turns a more or less bright pink. The laccase unit of measure is the amount of enzyme able to oxidise one nanomole of syringaldazine per minute, i.e.:

change in O.D. (at 530 nm) per minute x 46,15

We tested one in every 25 samples. About 2500 wines were submitted for small-scale distillation. The results of these analyses are given in the histogram shown in Figure 1. Analysis of the samples showed laccase units varying between zero and eleven.

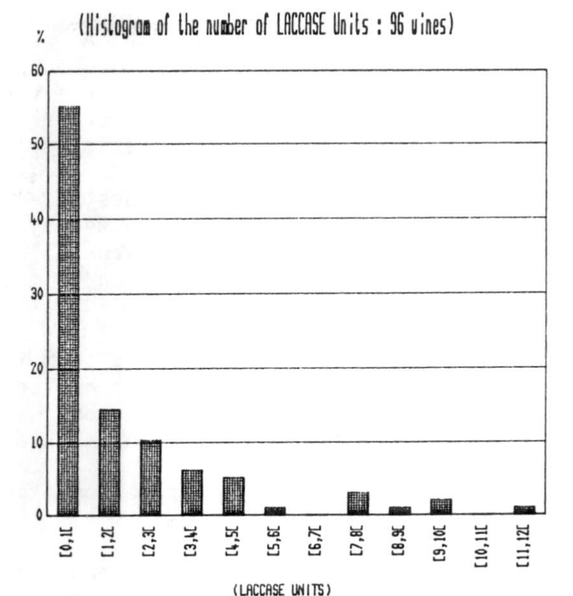

FIGURE 1 : Laccase test on 1987 vintages wines

According to the work carried out on wines for direct consumption (not for distillation) by the Institut d'Oenologie de Bordeaux by Grassin /1/ and Grassin et al /2/, problems did not arise with a laccase unit rate lower than two. Between two and three there was a possibility that the quality of the wine would be adversely affected (oxidasic casse) and above three the quality of the wine definitely deteriorated. We do not at present know whether this same scale would apply to wines used for distillation.

Nearly 70 % of the wines gave results of less than two laccase units, 10 % between two and three, 20 % more than three and only 4 % greater than seven.

In one of the tests conducted on the production of a large vineyard, the following relation was found :

Percentage of rot	laccase units
0	0
12	1,38
24	1,5
34	9,2

The weather in 1987 favoured the growth of **Botrytis cinerea**, particularly in cases where vines were poorly protected. This test also enabled us to ascertain that the level of contamination was relatively low.

Laccase activity decreases over a period of time. Wines distilled in March for example showed practically no laccase activity even when they derived from vintages which had been heavily botrytised. This test is therefore only indicative when conducted early on in the distillation season.

Tastings and more detailed analyses of spirits made from wines and must which were highly affected by **Botrytis cinerea** confirmed that organoleptic alterations already noted in wines by Cordonnier /3/ continue after ageing, namely :

- loss of fruitiness
- instability of aromas arising from fermentation
- appearance of camphor, phenol or iodine like tastes
- appearance of a lactone, sotolon, with a note of honey, sugar or caramel (4,5-dimethyl-3-hydroxy-2(5H)-furanone): Masuda et al /4/
- formation of flavours typical of oxidation (heavy prune, cooked or maderised notes).

With **Botrytis cinerea** it was noted that newly-distilled spirits were richer in ethyl acetate, ethyl butyrate and the C6 to C10 ethyl esters.

II - SMALL-SCALE DISTILLATION OF WINES FOR ORGANOLEPTIC EXAMINATION OF THE DISTILLATE

Background

Due to the fact that the defect of acrolein does not occur in wines because of its combination with polyphenols, the idea arose to carry out small-scale distillation of wine in a laboratory. When heated in an acid environment acrolein can be released and detected by organoleptic means.

The initial objective was to find a rapid test (brouillis + second distillation) to detect major defects in wines and thereby in the spirits into which they are distilled.

The hydraulic systems on the first grape-picking machines appearing on the market some twelve years ago were sometimes not sturdy enough and vintages were not always properly protected from this. When hydrocarbon pollution occurs due to an oil leak, small-scale distillation can be used to check whether wine of a particular vintage has been polluted or not.

In 1980, following freezing temperatures during the vintage season, this technique was frequently used to detect the well-known defects which occur when wine is made with grapes which have frozen (odour of pressed marc). Since then there has been an increased demand for small-scale distillation particulary from brokers and distillers.

Equipment and method

The equipment is made of glass (Figure 2). Initially copper turnings were introduced in the gooseneck tube and the condenser, but the distillates resembled those obtained for conventional analytic tests (strength for example). The copper quickly became covered with sulfides, mercaptans, sulfur-containing amino acids, fatty acids and, as a result, became inactive.

Several copper salts were tried : copper sulfate, copper nitrate and copper chloride, since in the pot stills the copper is uncoated and red (ionised copper : highly reactive cuprous oxide). Inexpensive copper sulfate provided good results.

Several tests were carried out with regard to the shape of the equipment :

- A round-bottomed flask heated by a Bunsen burner was tested, but the heat was poorly distributed.

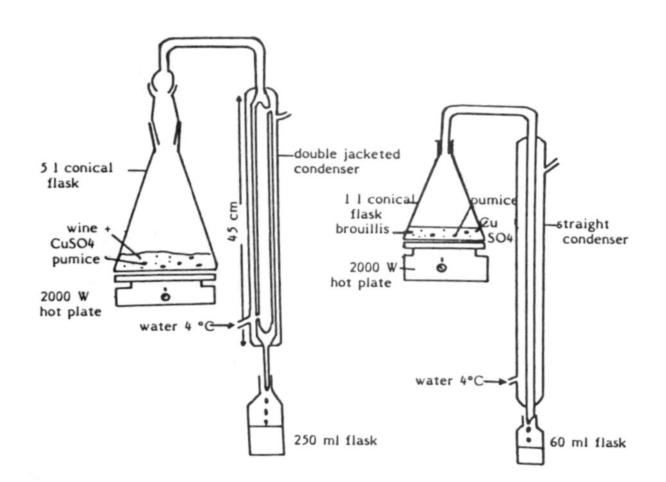

FIGURE 2 : Laboratory equipment for small-scale distillation

- With a 5 litre conical flask, organoleptic features of the distillate changed making it possible to detect major flaws in the wines by tasting. But a distillate identical to that from pot stills was never obtained. With pot stills the copper at certain spots reaches very high temperatures (500 to 600°C). With small-scale distillation the average temperature of the wine never exceeds 100°C.

To obtain a sufficiently cooled distillate (near the ambient temperature), especially if several apparatuses have been connected in series, a double jacketed condenser and a low-temperature cooling liquid (about 4°C) are necessary to distill the wine.

To obtain systematically 250 ml of a 27 % vol brouillis, the volume of wine to be used will depend on its alcoholic strengh. For example : 1025 ml of wine at 7 % vol would be used, whereas 779 ml of wine at 9 % vol would be used.

1 - Brouillis

Distilling the wine requires several drops of an organoleptically neutral antifoam agent (Prolabo : Rhodorsil silicone), five to seven grams of copper sulfate and several chips of pumice to stabilize the boiling.

2 - Second Distillation

The 250 ml of 27 % vol. brouillis is distilled in the presence of 2 to 3 grams of copper sulfate and chips of pumice. The strength of the resulting spirit is about 70 to 71 % vol.

Distilling the wine takes about twenty to twenty-five minutes, that of the brouillis from five to ten minutes. It must be performed quickly since the large headspace of the laboratory equipment causes more cooling of the vapour and hence greater risk of rectification and distillation of seconds likely to mask certain defects.

As with pot stills, the sulfides are fixed by copper sulfate. We carried out tests to determine the optimum quantity of copper sulfate since it can partially fix butyric acid but does not react with ethyl butyrate.

About 3000 microdistillations are performed yearly at the Station Viticole. They are also performed at other regional laboratories.

The main defects in wine we encounter are :
- hydrocarbon (pollution subsequent to oil leaks)
- acrolein
- putrid (butyric)
- sourness
- bitterness
- oxidised, maderised

When such defects are detected the wine grower can eliminate the spoiled wine to maintain the quality of his other spirits. Oenologists may then search for causes and give the requisite advice.

III - DISTILLATION

The distillation period covers several months from the end of the fermentation (begining of November) up to March 31. The traditional Charentais distillation process uses a pot still such as that shown in Figure 3 and is carried out in two stages: brouillis and a second distillation. It has been described in detail by Lafon et al /5/. Two methods are used. In the first method the seconds are recycled back to the wines, in the other back to the brouillis. The pot is no larger than 30 hectolitres including the 25 hectoliter load. Caumeil /6/ has described the various fundamentals of the procedure which are summarised in Table 1.

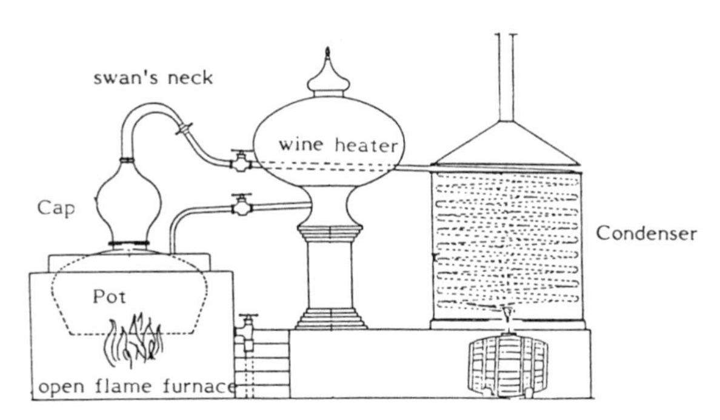

FIGURE 3 : Charentais pot still

TABLE 1 : Example of the procedures involved in the two main cognac distillation methods

a) Recycling seconds back into the wines

<u>Distillation of the wine</u>

Wine at 9 % vol (2250 l)
+ seconds at 30%vol (230)
+ heads at 67% vol (20 l)

= 2500 l at 11% vol —Dist→

heads at 57%V (10 l)
brouillis at 30%V (920 l)

<u>Distillation of the brouillis</u> (second distillation)

Brouillis at 30 % vol (2500 l) — Dist →

heads at 78 % vol (25 l)
seconds at 30 % vol (700 l)
spirit at 70 % vol (730 l)

b) Recycling seconds back into the brouillis

<u>Distillation of the wine</u>

Wine at 9 % vol (2320 l)
+ heads at 65 % vol (20 l)
+ tails at 3 % vol (160 l)

= 2500 l at 9% —Dist→

heads at 55%V(10 l)
brouillis at 27%V (790 l)
tails at 3%V (120)

<u>Distillation of the brouillis</u> (second distillation)

2/3 brouillis at 27 % vol
+ 1/3 seconds at 30 % vol

= 2500 l at 28%vol —Dist→

heads at 76%vol (25 l)
spirit at 70 % vol (680 l)
seconds at 30 % vol (650 l)
tails at 3 % vol (80 l)

The maximum alcoholic strength of the distillaton is 72 % vol. Heating is carried out over an open flame. Tests were conducted with dual energy sources.
. flow started by gas
. flow by electricity (tubular heating rods or heating elements on both sides of the burner).

The consumption of electricity expressed in terms of a hectoliter of pure alcohol produced is about 88 % compared with the same amount distilled with gas. Considering the cost of modifying the stills and the French Electricity Authority's (EDF) expensive rates at a time of the year when consumption of electricity is high, this gain in energy was not deemed sufficient and consequently the tests were halted.

The first wines distilled have not necessarily undergone malo-lactic fermentation. This is not a problem ; however, it is recommended not to distill the wine when malo-lactic fermentation is proceeding.

Over the five-month distillation period, the composition of wines varied. Polyol and higher alcohol concentrations were unchanged. However, the organoleptically important esters (isoamyl, hexyl and phenylethyl acetates and ethyl caproate, caprylate, caprate and laurate) concentrations decreased significantly and ethyl acetate, ethyl lactate, diethyl succinate, acetaldehyde and acetic acid increased as shown in Table 2.

The volume of heads removed during the distillation is therefore not the same at the beginning and end of the distillation period (approximately 1,5 % in the beginning, less than 1 % at the end). In order to avoid coarser elements appearing in the spirit, the wines are generally racked in order to eliminate most of the less.

In accordance with tradition and local custom, distillation is carried out with natural wine on its fine lees. Of course each cognac house provides its distillers with special instructions in order to obtain the specific type of product they each market.

Such instructions mainly concern the proportion of fine lees to be kept in the wines used. Obviously this results in considerable analytical differences in the final product.

The presence of lees results in a higher quantity of organoleptically important esters and produces spirits which are richer in floral aromas (Table 3). Care must be taken that these lees are of good quality, otherwise they are a major source of defects.

TABLE 2 : Changes in the concentrations of consituents of wine conserved over a five-month period

Constituents		% change
Ethyl acetate		+ 24,1
Isoamyl acetate	1	- 55,7
Hexyl acetate	2	- 60,7
Phenylethyl acetate	3	- 62,1
Ethyl caproate	4	- 22,0
Ethyl caprylate	5	- 14,2
Ethyl caprate	6	- 18,9
Ethyl laurate	7	- 52,0
Sum of esters	1 to 7	- 25,8
Ethyl lactate		+ 49,1
Diethyl succinate		+ 226
Acetaldehyde		+ 117
Acetic acid		+ 18,8

TABLE 3 : Effect of lees on ester content

Constitutents (mg/l spirit at 70 % vol)	Distillation with smaller proportion of lees	Distillation with lees
Ethyl caproate	6,76	8,3
Ethyl caprylate	8,95	23,6
Ethyl caprate	13,8	63,0
Ethyl laurate	12,45	36,2
Ethyl myristate	5,4	9,8
Ethyl palmitate	9,77	13,2
Ethyl palmitoleate	1,44	1,8
Ethyl stearate	0,59	0,61
Ethyl oleate	1,19	1,22
Ethyl linoleate	7,69	9,52
Ethyl linolenate	1,86	2,58
Isoamyl caprylate	0,42	2,48
Isoamyl caprate	1,67	5,76
Isoamyl laurate	0,78	1,83
2-Phenylethyl caprylate	traces	1,20
2-Phenylethyl caprate	0,25	1,55
Total esters	73,02	182,65(=+150%)

Wine aromas, whether volatile or carried over with the aqueous-alcohol vapours, do not all pass over at the same speed during distillation. Figure 4 shows typical graphs for 39 constituents obtained during wine distillation (to obtain the brouillis) and brouillis distillation (the second distillation). In the graphs of the second stage of the distillation, the areas corresponding to head and main "heart" fractions are given.

Brouillis : <u>Type 1</u> : acetaldehyde, 1,1-diethoxyethane, 1,1-diethoxy-2-methylpropane, ethyl acetate, propionate and butyrate, ethyl caproate, caprylate, caprate, laurate, myristate, palmitate, stearate, oleate, linoleate and linolenate, isoamyl acetate, isoamyl caprate, pH

<u>Type 2</u> : furfural

<u>Type 3</u> : methanol

<u>Type 4</u> : 1-propanol, isobutanol, 1-butanol, 2-methyl and 3-methyl-1-butanol, phenylethyl acetate

<u>Type 5</u> : 2-phenylethanol

<u>Type 6</u> : ethyl lactate and diethyl succinate

Second distillation

<u>Type 1</u> : acetaldehyde, 1,1-diethoxyethane, 1,1-die thoxy-2-methylpropane, ethyl acetate, propionate and butyrate, hexyl acetate, ethyl caproate, laurate, myristate and palmitate, isobutyl caprate, isoamyl acetate, isoamyl caprylate, caprate and myristate

<u>Type 2</u> : 2-phenylethanol

<u>Type 3</u> : methanol

<u>Type 6</u> : furfural, phenylethyl acetate, ethyl lactate, diethyl succinate, caprylic, capric and lauric acids, total acidity

<u>Type 7</u> : 1-propanol, isobutanol, 1-butanol, 2-methyl and 3-methyl-1-butanol, ethyl stearate, oleate, linoleate and linolenate, pH

<u>Type 8</u> : ethyl caprylate, ethyl caprate

These various graphs help to understand the difficulties involved in distillation. To obtain high quality spirit requires a balance among all the constituents and the distiller's art is necessary to achieve such a balance.

Brouillis (First distillation)

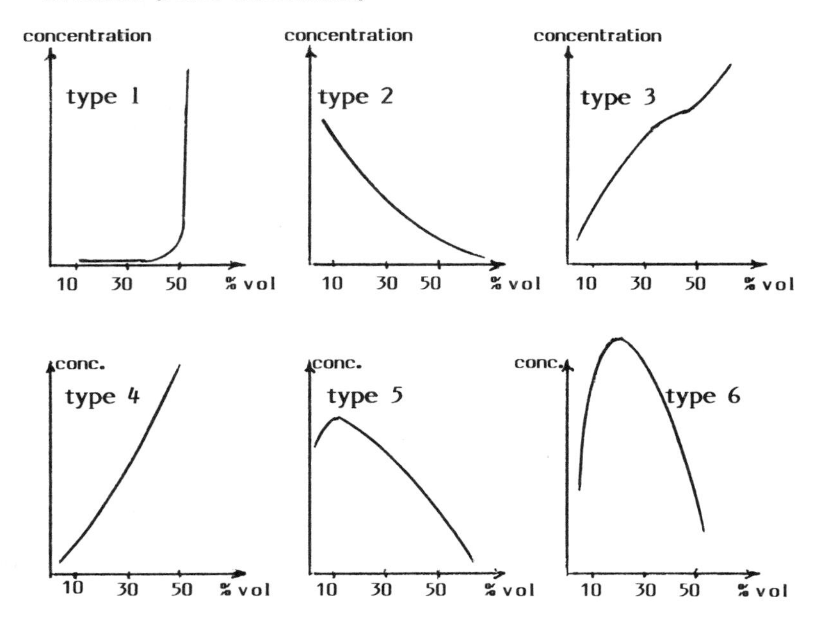

Bonne chauffe (second distillation)

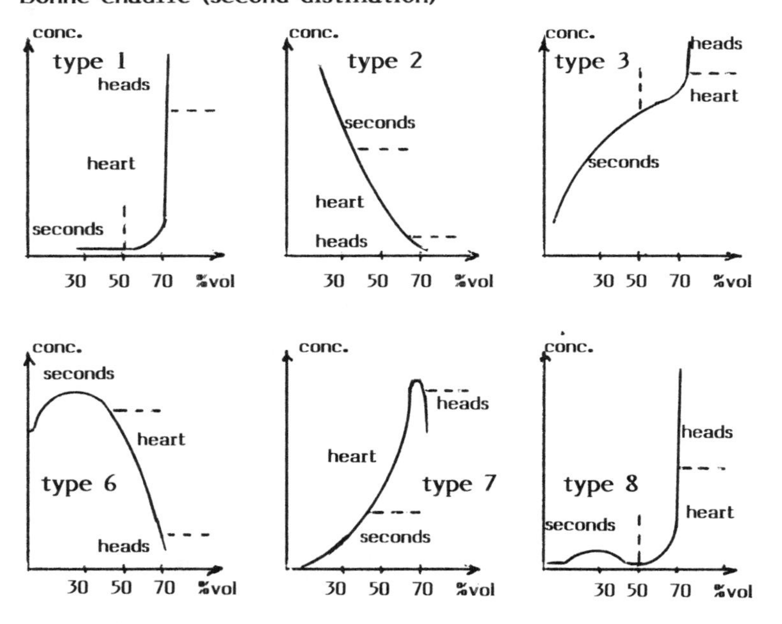

FIGURE 4 : The passing over of constituents during distillation

IV - GREATER ANALYTICAL KNOWLEDGE OF COGNACS

Although verifications of "appellation d'origine controlée" (designated origin) wines and spirits by organoleptic means was first officially formalised then made mandatory by French national and local regulations it has, in fact, always been performed by dealers and cellar masters. There are now various procedures for qualitative examinations, but the tastings are all performed by the same people ; hence the economic and psychological impact remains consistent.

Protecting the product's high-quality image requires a method of chemical analysis able to furnish greater knowledge of cognac, give a detailed picture of its composition, and thus provide a supplementary guarantee of its origin and organoleptic quality.

Ensuring the best product quality requires ascertaining defects and their causes so that an oenologist can give the technical advice to eliminate or avoid repetition of such defects. An in-depth analysis (about one hundred parameters) of course provides a great deal of data, but is difficult to carry out on a large number of samples.

We focused on a simpler analytical method which detects certain defects and is applicable both to new spirits and to the finished product : gas chromotography of the spirits by direct injection for some of the volatile compounds. The number of samples to be tasted is reduced accordingly.

These analyses can ascertain the following defects :

- oxidation, maderisation : acetaldehyde, acetal (1,1-diethoxyethane)
- sourness, acescence : ethyl acetate
- stagnant and butyric : 1-butanol, 2-butanol, ethyl butyrate
- burnt plastic : acrolein, allyl alcohol
- fermentation problems if the propanol contents are higher than that of isobutanol, acetoin
- poor distillation (seconds, tails) : excess ethyl lactate and 2-phenylethanol
- excess heads : ethyl esters (C8, C10, C12), aldehydes and higher alcohols

The tolerance thresholds for the different compounds mentioned can be determined by the following steps :

- analysing spirits showing defects compared with good quality spirits ;

- tasting good quality spirits to which various corresponding chemical compounds have been added.

Problems sometimes arise with this method of adding compounds since it changes the pre-existing chemical balance. The perception of the compound is not the same if the tasting takes place soon after the addition has been made compared with several days later.

In figure 5, two gas chromatograms are shown, one of a clearly defective spirit produced from a wine having undergone substantial microbial spoilage, the other of a good quality spirit. Table 4 summarizes the analytical differences noted.

Initial tests on new spirits distilled in 1987-88 indicated the following taste thresholds, above which the spirits were judged to be defective :

60	mg/l	for acetaldehyde
30	mg/l	for acetal
600	mg/l	for ethyl acetate
6	mg/l	for 1-butanol
6-7	mg/l	for 2-butanol
4-5	mg/l	for ethyl butyrate

TABLE 4 : The chemical composition of a good quality spirit compared with that of a spirit made from wine having undergone substantial bacterial spoilage

mg/l of spirit at 70° vol	Good quality	Bacterial spoilage
Acetaldehyde	25,5	21
1,1-Diethoxyethane	9,45	8,92
Methanol	446	532
Propanol	269	287
Isobutanol	936	1375
1-butanol	1,80	14,8
2-butanol	1,84	23,6
2-methyl-1-butanol	408	577
3-methyl-1-butanol	1708	2212
1-hexanol	20,5	13,8
cis-3-hexen-1-ol	2,78	2,70
Allyl alcohol	0,9	8,22
Ethyl formate	9,1	44,8
Ethyl acetate	234	306
Ethyl butyrate	0,75	28,2
Ethyl lactate	153	422

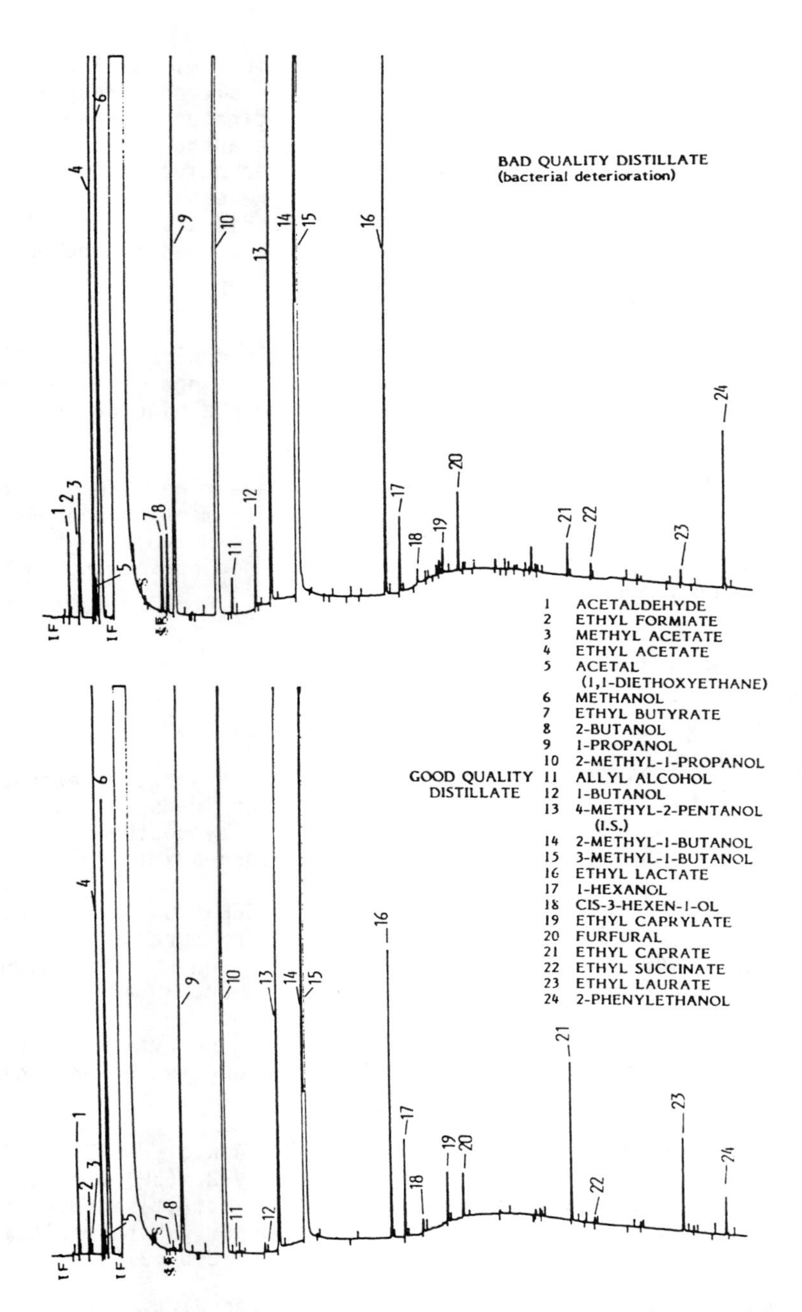

FIG. 5 : Typical chromatograms of a good quality spirit and of a spirit produced from wine having undergone spoilage

We noted that over a greater number of samples major positive correlations were noted between C4 alcohols, ethyl butyrate and ethyl lactate. Ethyl lactate indicates whether the wines were distilled early or late during the authorised period. Since the wines are not protected by preservatives, in order to prevent such bacterial spoilage, winegrowers must not be negligent regarding the application of fungicides and special care should be taken to ensure the cleanliness of winemaking equipment.

CONCLUSION

The development of analytical techniques and their close relationship with results from tastings give reason to expect major progress to be made in ascertaining defects and quality factors in spirits.

Much work remains to be done to determine with greater accuracy the organoleptic flavour thresholds mentioned above as well as those for many other constituents, particularly by checking and comparing them over several years and performing the same tests on finished products.

REFERENCES

1 - GRASSIN C.; Recherches sur les enzymes extracellulaires secretées par Botrytis cinerea dans la baie de raisin. Applications oenologiques et phytopathologiques. Doctoral thesis for the Université de Bordeaux II, n° 28, June 1987

2 - GRASSIN C.; DUBOURDIEU D., BERNARD JL.; Prévision des conséquences néfastes de la pourriture grise sur la qualité des vins : l'estimation des risques d'oxydation. Phytoma, Défenses des cultures n° 389, 1987, 39-42

3 - CORDONNIER R.; Les effets de Botrytis cinerea sur la couleur et l'arôme. Revue française d'oenologie, n° 108, 1987, 21-29

4 - MASUDA M.; OKAWA E.; NISHIMURA K.; YUNOME H.; Identification of 4,5-dimethyl-3-hydroxy-2 (5H)-furanone (Sotolon) and ethyl 9-hydroxynonanoate in botrytised wine and evaluation of the roles of compounds characteristic of it. Agric. Biol. chem., 48(11) 1984, 2707-2710

5 - LAFON J.; COUILLAUD P.; GAY-BELLILE F.; Le Cognac, sa distillation, 5th edition. Editions Baillière Paris

6 - CAUMEIL M. Le Cognac. Pour la Science, Dec 1983, 48-57

11

Fruit distillate flavours

W. Postel and L. Adam
Institut für Lebensmitteltechnologie und Analytische Chemie, Technische Universität München, D-8050 Freising-Weihenstephen, West Germany

The flavour of fruit distillates has already been discussed in a monograph by Nykänen and Suomalainen in 1983 /1/ and in a survey by Ter Heide in 1986 /2/. The genuine fruit flavours and their relevance to the flavour of the final distilled beverage have been reviewed by Dürr and Tanner in 1983 /3/. Lists of volatile compounds in food that include data about fruit brandies have been continually published by Maarse and Visscher since 1983 /4/. In the last few years investigations to quantify the volatile components of fruit distillate flavours have been intensified in order to obtain better knowledge of the authenticity, purity and quality of these spirits, on the state and quality of raw materials used, on the processing procedures and on possible adulterations. This paper deals with those quantitative results obtained with brandies of stone fruits (plums, mirabelles, cherries), pome fruits (pome, Williams pears, Calvados), pomace, and raspberries.

Fruit distillates differ widely in quality and composition, dependent on the raw material used and the processing procedures. They are especially rich in volatiles. In comparison with other spirit groups they contain the highest amounts of alcohols and esters (Fig. 1). The sum of volatiles in fruit brandies amounts to an average of nearly 2000 mg/100 ml ethanol with the exception of calvados and raspberry brandy. Raspberry brandy is a distillation product of unfermented raspberry fruits with the addition of neutral alcohol (Germany) or of alcohol originating from other fruits (France). Calvados is a French distillation product of cider. The other fruit brandies from plums, mirabelles (small yellow plums), cherries, pears and apples are usually derived from a fermented mash as well as pomace distillates.

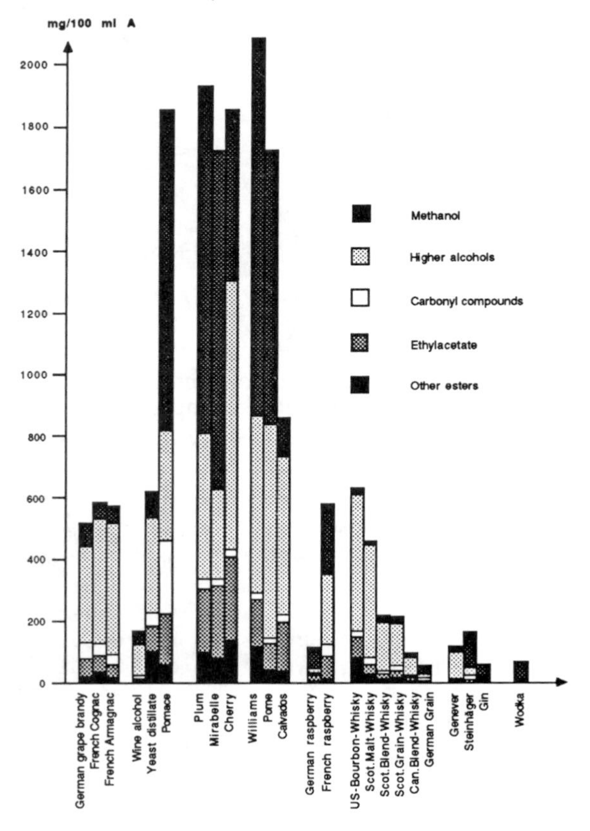

Figure 1- Volatiles in brandies

STONE AND POME FRUIT BRANDIES

Methanol

Methanol is quantitatively the main component of stone and pome fruit brandies, excluding Calvados (Fig. 1). As a rule plum-, mirabelle- and Williams-distillates contain more than 1000 mg/100 ml ethanol, whereas cherry distillates range lower, from 400 - 800 mg/100 ml ethanol /5/. Since during fermentation of the fruit mash a certain minimum amount of methanol is formed by enzymatic cleavage of pectin, up to now the methanol content of fruit brandies has been used for the evaluation of authenticity and possible adulterations such as addition of neutral alcohol or of distillates from cheap materials or addition of sugar to the mash. Cherry distillates for example are expected to contain a methanol concentration of at least 400 mg/100 ml ethanol /6, 7/. Similar rules exist for the other stone and pome fruit brandies /8/.

With regard to the toxicological effects of methanol a low concentration is desirable. Therefore several ways to decrease the methanol content of fruit brandies have been discussed /9, 8/, such as heat treatment of the mash to inactivate the pectolytic enzymes or fruit juice fermentation instead of mash fermentation, but there are still technological and sensory problems /5/. The removal of methanol from the brandy by a special distillation procedure or from the enzym treated mash before fermentation is also possible /8/, but results in a loss of flavour volatiles /3/.

Higher alcohols, terpenes, carbonyl compounds

In comparison with other spirit groups, fruit brandies contain high amounts of 1-propanol, 1-butanol, 2-butanol and 1-hexanol. Besides fruit brandies, only grape wine brandies have amounts of 1-hexanol that are worth mentioning /10/. Isobutanol and the two isoamylalcohols are main components of fruit brandies, but their concentrations are approximately in the same range as in other spirit groups such as whiskies and grape brandies. Other aliphatic alcohols that are present in smaller amounts include allylalcohol, 1-pentanol, 3-methyl-1-pentanol, cis-3-hexen-1-ol, trans-2-hexen-1-ol, 1-octanol, 1-decanol and 1-dodecanol. Benzylalcohol and 2-phenylalcohol are the most important aromatic alcohols /10/.

Some terpene compounds, such as α-terpineol, geraniol, linalool, cis- and trans-linalooloxid have also been quantitatively determined in fruit brandies. As a rule the concentrations of the individual components are lower than 1 mg/100 ml ethanol /5, 10, 13/.

Among the carbonyl compounds acetaldehyde and 1,1-diethoxyethane are dominant; the mean values range from 9 to 17 and from 4,5 to 9,5 mg/100 ml ethanol, respectively /5/. Other carbonyl compounds present in fruit brandies are propionaldehyde, isobutyraldehyde, acrolein, benzaldehyde, furfural, acetone, methylethylketone, acetoin and 1,1,3-triethoxypropane and some others in minor amounts /5, 11/.

There are marked differences between stone and pome fruit distillates /5, 10, 12-15/. Stone fruit distillates are characterized by relatively high amounts of benzylalcohol and benzaldehyde, pome fruit distillates by high amounts of

1-hexanol. As a rule the terpenes were found in higher concentrations in stone fruit brandies than in pome fruit brandies (Fig. 2) /5/.

Benzaldehyde as well as prussic acid are formed during mashing and fermentation through hydrolysis of amygdalin present in the fruit stones; benzylalcohol is produced through reduction of benzaldehyde during fermentation. The final concentrations of these components depend on technological factors like quantity of crushed stones, acidification of the mash, enzyme treatment and storage time of the mash /3/. Furthermore the amount of benzylalcohol in stone fruit brandies depends considerably on the distillation procedure. The same is true for 2-phenylethanol. In a two step batch distillation for example the final distillate contained only 3% of benzylalcohol and 4% of 2-phenylethanol present in the mash /5/. Thus, the concentrations of these two aromatic alcohols are less suited as quality indicating compounds, but they give indications about the distillation procedure.

The average content of 2-butanol in fruit brandies is very high /16, 5/. The highest amounts have been found in Calvados, pome and Williams distillates (Fig. 2). In each fruit brandy group there are products with very low and very high concentrations. This 2-butanol content depends on the activity of Lactobacilli strains in the fermented mash during storage /17, 5/ (Table 1). Allylalcohol, acrolein, acetoin and 1,1,3-triethoxypropane as well as excessive amounts of 1-propanol, ethyl-acetate, -propinate, -succinate, and -lactate are also due to bacterial activities in the fermented mashes or wines. Thus, these substances are helpful factors in the evaluation of the quality of the mashes used /5/.

Table 1 - Influence of storage of the fermented mash on the concentrations of 1-propanol and 2-butanol (mg/100 ml ethanol) /5/

	Distillation	
	immediately after fermentation	after a storage time of 4 months
1-propanol	80 - 100	510 - 1450
2-butanol	not detectable	30 - 95

A characteristic of cherry distillates is its low content of 1-butanol (Fig. 2). The mean value of 64 samples was 2 mg/100 ml ethanol, the highest concentration was 5 mg/100 ml ethanol /22/. A further characteristic of cherry distillates is its extremely high concentration of 1-propanol /10, 6, 5/, which ranges in commercial products from 478 to 2225 mg/100 ml ethanol /10/. However, in cases where the mashes were distilled immediately after fermentation the concentrations

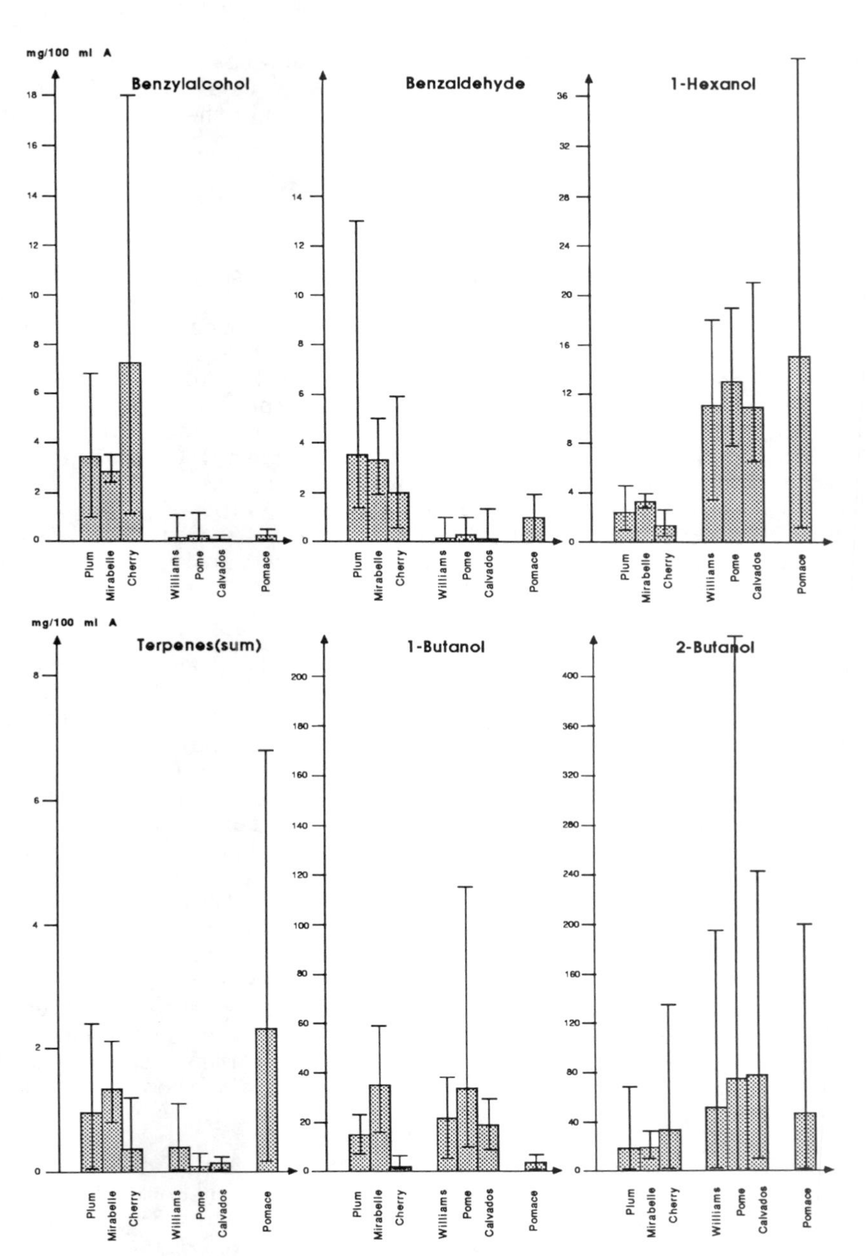

Figure 2 - Benzylalcohol,benzaldehyde,1-hexanol,terpenes(sum),1-butanol and 2-butanol in fruit brandies (mean and range)

were only 80-100 mg/100 ml ethanol (Table 1). The longer the fermented mashes were stored, the higher the concentrations of 1-propanol due to bacterial activity became /5/. Extremely high amounts of 1-propanol, which are considered to decrease the quality of the spirit /6/, were only formed in the fermented mashes of cherries, in other fruit mashes the formation was much lower.

Esters

The main ester component of fruit brandies is ethyl acetate followed by ethyl lactate; together these two compounds amount to about 80% or more of the total ester content. The number of the other esters is large, but their concentrations are relatively small. Most of the esters are ethyl esters beginning with formate up to palmitate, phenylacetate, benzoate and cinnamate, including some hydroxy esters. The number of isoamyl and methyl esters is smaller; in addition there are propyl, butyl, hexyl, 2-phenethyl and benzyl esters, mainly acetates /10, 6, 5, 13/. Worth mentioning is methyl acetate, because fruit brandies as well as pomace distillates are the only spirit groups with higher levels of this ester; in grape wine brandies and whiskies it occurs only in traces /5, 10/.

A marked distinction between stone and pome fruit distillates exists in the concentrations of ethyl benzoate, diethyl succinate and benzyl acetate (Fig. 3); these compounds occur in much higher concentrations in stone than in pome fruit spirits /10, 5/. The latter contain relatively high amounts of hexyl acetate. Cherry spirits are usually characterized by high concentrations of propyl acetate (up to 38 mg/100 ml ethanol) and benzyl acetate (up to 1,1 mg/100 ml ethanol). In Calvados markedly higher amounts of 2-phenethyl acetate (up to 7.4 mg/100 ml ethanol) than in other fruit brandies are found /5/.

The flavours of plum-, mirabelle-, cherry- and apple-brandies as well as the flavour of pomace brandies are not based on character impact compounds. The individual flavour is assumed to depend on the balance of many components deriving partly from the original fruit, partly from the fermentation and processing procedure /18, 19, 11, 2, 3/.

The most conspicuous fruit brandy with regard to its composition is Williams pear brandy. Williams distillates are the only ones with high amounts of ethyl-3-hydroxy octanoate and of many unsaturated esters. Ter Heide /2/ listed 24 unsaturated compounds /20, 21, 10, 5/. Quantitative results of some unsaturated esters are given in Table 2 /22/.

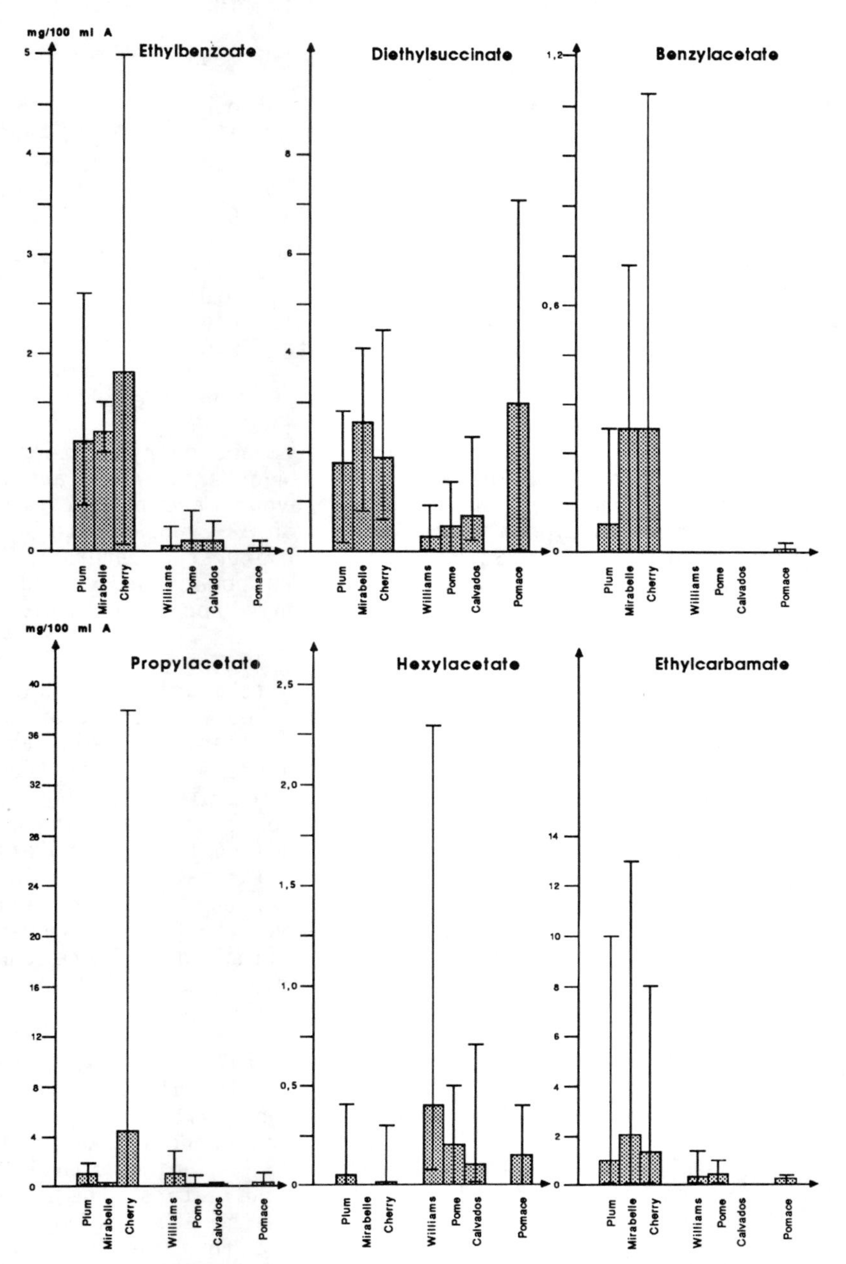

Figure 3 - Ethylbenzoate,diethylsuccinate,benzylacetate,propylacetate,hexylacetate and ethylcarbamate in fruit brandies (mean and range)

Table 2 - Esters of unsaturated fatty acids in 26 commercial Williams pear brandies (mg/100 ml ethanol) /22/

	mean	range
Ethyl-trans-2-octenoate	0,09	+ - 0,4
-cis-4-decenoate	1,4	0,7 - 2,5
-trans-2-decenoate	0,4	n.d. - 2,1
-trans-2-cis-4-decadienoate	18	0,8 - 61
-trans-2-trans-4-decadienoate	5	0,5 - 15
Methyl-cis-4-decenoate	0,05	+ - 0,15
-trans-2-decenoate	0,01	+ - 0,08
-trans-2-cis-4-decadienoate	6,6	2,6 - 19
-trans-2-trans-4-decadienoate	2,0	0,9 - 5,2

+ = < 0,01 mg/100 ml ethanol; **n.d.** = none detected

Main components are the ethyl and methyl esters of the decadienoic acid. The trans-2-cis-4-esters, which are responsible for the typical Williams flavour, are partly isomerized to the non-typical smelling trans-2-trans-4-esters during fermentation /3/. Some of these unsaturated esters were also detectable in small amounts in a few pome brandies, whose raw materials are known to be not only apples but also to a certain extent pears /5/.

The ester concentrations of fruit brandies are influenced by technological factors. As a rule heat treatment of the mash increases the contents of higher esters as well as juice fermentation /5/. The storage of the fermented mash leads also to an increase of most esters /5/. The distillation procedure can also affect the ester composition of spirits. The majority of esters are heads components. Heads cutting during distillation to reduce acetaldehyde and ethylacetate results in losses of higher fatty acid esters /23-25/. On the other hand ethyl lactate and diethyl succinate are extreme tails components. In a two step batch distillation the final distillate contained only 9 to 14% each of ethyl lactate and diethylsuccinate present in the fermented fruit mash /5/.

The greatest losses of esters take place when reducing the distillates with water from 65 - 70% vol to 40 - 45% vol alcohol and the following chilling and filtration process. The lower esters as well as alcohols and carbonyl compounds were not affected, but the higher esters beginning with hexanoate or octanoate decreased; the higher the esters, the more these levels decreased up to nearly 100% (e.g. ethyl palmitate). The losses of unsaturated Williams esters for example amounted to between 30 and 70% /5/. Since these esters are aroma impact substances valuable flavour components are inevitably removed by the unavoidable chilling and filtration procedure.

Ethyl carbamate (Urethan)

Towards the end of 1985 high concentrations of ethyl carbamate, a known animal carcinogen, have been detected in alcoholic beverages, above all in fruit brandies, by Canadian chemists /26/. The results have been confirmed by investigations in Europe /27-31/. The highest amounts up to 20 mg/l have been found in stone fruit brandies. As an example the results of commercial products analysed in 1986 are shown in Figure 3 /32/.

Ethyl carbamate has already been detected a decade ago in beverages treated with diethyl pyrocarbonate; besides, it has been demonstrated that it occured naturally in fermented foodstuffs such as wine, beer, bread, yoghurt, soy sauce and sake, but only at µg/kg levels, that were postulated to originate from carbamyl phosphate /33/.

The formation of those high levels of ethyl carbamate in fruit brandies has been intensively investigated and discussed in the last two years /28, 34-39/. Up to now the results have lead to the assumption that the formation depends on the concentrations of prussic acid, deriving from amygdalin, and 1,2-dicarbonyl compounds, such as diacetyl, 2,3-pentanedion and methyl glyoxal. The concentrations of alcohol and oxygen and the pH value are of importance too. In addition, the formation of ethyl carbamate highly depends on the influence of light; the maximum acceleration of the reaction is in the spectral range from 330 to 340 nm. The reaction takes mainly place in the distillate. Distillates that are kept completely dark formed only negligible amounts of ethyl carbamate or none at all.

For reasons of health the Canadian authorities established the following guideline limits for the presence of ethyl carbamate in alcoholic beverages: 30 µg/l for wine, 100 µg/l for fortified wines, 150 µg/l for liqueurs, and 400 µg/l for distilled spirits /38/.

The most affected alcoholic beverages are stone fruit brandies (Fig. 3). Following the Canadian guidelines, the German Federal Health Office set an upper limit of 0,4 mg/l ethyl carbamate for fruit spirits. The producers of fruit brandies had to face a lot of difficulties in the past two years to comply with this limit, because there is so far no satisfactory solution for this problem. Some companies have even ceased producing stone fruit brandies. The knowledge of the formation of ethyl carbamate and its prevention is still insufficient.

So far there are only several suggestions about appropriate ways to avoid high ethyl carbamate concentrations

in fruit brandies, but none of them is fully satisfactory.
The longer the storage of the fermented mash, the more diacetyl may be formed by microbial activity. Hence, the fermented mash should not be stored for several months, it should be distilled without delay, at the latest after several weeks. But then, as a rule those distillates have only a faint typical flavour.
The removal of prussic acid from the mash or distillate by precipitation with copper chloride may also be a way to reduce the ethyl carbamate concentration in the final spirit. Apart from the environmental problems with the high copper contents in the distillation residues, there is up to now little experience with this process.
Ethyl carbamate that has already formed in a spirit can be removed to a large extent by redistillation and early tails cutting since ethyl carbamate is an extreme tails component. But if there are still precursers in the distillate, ethyl carbamate will again be formed under the influence of light. So far it seems to be a real problem to avoid high concentrations of ethyl carbamate in fruit brandies as well as to achieve a high quality spirit.

POMACE BRANDIES

Pomace brandies are made by fermenting sugars left in residues of the grape pressing procedure in the wine making process. The distillates obtained are called Grappa in Italy, Marc in France and Tresterbranntwein in Germany. As shown in Figure 1 the composition of pomace distillates is similar to fruit brandy distillates with the exception of carbonyl compounds, mainly acetaldehyde and 1,1-diethoxyethane, which are present in much higher amounts in pomace distillates.
In general pomace distillates are also characterized by relatively high amounts of certain alcohols such as methanol, 1- and 2-propanol, 2-butanol and 1-hexanol (Fig. 2), of acrolein, hexanal, acetoin, and of some esters such as ethyl acetate, -lactate, -butyrate, diethyl succinate (Fig. 3), methyl-, propyl-, butyl-, isobutyl- and 2-phenethylacetate /25/. However, the determination of 75 compounds in two German, three Italian and eight French pomace brandies showed, that there were considerable differences between the several products (Table 3). For example, in French pomace brandies ethyl acetate concentrations ranged from 8 to 397 mg/100 ml ethanol, acetaldehyde concentrations from 9.6 to 477 mg/100 ml ethanol. One of the German products (first column in Table 3) was remarkable for its low methanol and 1-propanol contents (322 and 20 mg/ 100 ml ethanol, respectively) and its extremely low concentrations of isobutanol and isoamylalcohols (6.9 and 12 mg/ 100 ml ethanol, respectively). This product was made by a special rectificating distillation procedure to reduce the concentration of methanol, but the sensory qualities were not very satisfactory.

Table 3 - Volatiles in pomace brandies (mg/100 ml ethanol)

	German(Trester)		Italian(Grappa)			French(Marc)		
Number of samples	2		3			8		
			range			range		
Methanol	322	955	427	-	832	415	-	1682
1-Propanol	20	61	53	-	80	43	-	91
2-Propanol	2,9	2,1	1,2	-	1,7	+	-	4,3
Allylalcohol	0,5	0,9	1,0	-	3,8	n.d.	-	1,8
1-Butanol	0,2	1,8	4,3	-	6,4	2,2	-	4,6
2-Butanol	32	51	27	-	200	0,9	-	100
2-Methyl-1-propanol	6,9	67	58	-	91	46	-	47
2-Methyl-1-butanol	2,8	29	36	-	75	29	-	62
3-Methyl-1-butanol	9,5	88	114	-	200	105	-	154
Isoamylalcohols, Sum	12	117	150	-	275	140	-	216
1-Pentanol	0,03	0,3	0,4	-	0,7	0,4	-	1,0
2-Pentanol	1,1	0,2	+	-	0,5	+	-	0,5
3-Methyl-1-pentanol	n.d.	n.d.	+	-	0,04	+	-	0,06
1-Hexanol	1,2	23	10	-	11	6,5	-	39
cis-3-Hexen-1-ol	0,06	0,7	0,8	-	1,0	0,4	-	2,3
trans-2-Hexen-1-ol	0,02	0,6	0,3	-	0,4	0,04	-	1,0
1-Heptanol	n.d.	n.d.	n.d.	-	n.d.	n.d.	-	0,6
2-Heptanol	n.d.	n.d.	+	-	0,02	0,03	-	0,2
1-Octanol		0,3	0,2	-	0,3	0,06	-	0,5
1-Nonanol	0,02	n.d.	n.d.	-	0,03	0,03	-	0,5
1-Decanol	n.d.	n.d.	0,1	-	0,1	n.d.	-	0,2
1-Dodecanol	0,02	n.d.	n.d.	-	0,03	+	-	0,02
2-Phenylethanol	0,1	0,6	1,5	-	6,0	1,1	-	3,2
Benzylalcohol	0,1	0,1	0,02	-	0,2	0,08	-	0,5
a-Terpineol	0,01	0,1	0,04	-	0,09	+	-	0,4
Geraniol	n.d.	0,04	0,05	-	0,06	0,03	-	2,5
Linalool		0,1	0,02	-	0,1	0,03	-	0,8
cis-Linalooloxid	0,08	0,9	0,11	-	0,14	0,05	-	1,8
trans-Linalooloxid	0,07	0,8	0,15	-	0,15	0,07	-	3,1
Acetaldehyde	20	21	59	-	134	9,6	-	477
Propionaldehyde	+	0,3	0,4	-	0,6	n.d.	-	1,0
Isobutyraldehyde + Aceton	1,0	2,2	1,0	-	5,8	0,3	-	8,1
Hexanal	1,0	2,4	1,0	-	1,2	n.d.	-	2,1
Heptanal	n.d.	+	+	-	0,04	n.d.	-	0,09
Acrolein	0,2	1,2	1,3	-	4,2	0,2	-	5,5
3-Hydroxybutan-2-one	0,1	0,3	1,5	-	2,5	n.d.	-	6,3
Benzaldehyde	+	1,5	0,9	-	1,0	0,5	-	1,9
Furfural	0,3	0,7	0,3	-	0,6	0,2	-	6,5
1,1-Diethoxyethane	8,6	14	18	-	72	8,5	-	135
1,1,3-Triethoxypropane	0,08	0,2	0,08	-	0,2	0,05	-	0,2
Ethyl -formate	6,3	+	n.d.	-	0,8	n.d.	-	1,5
-acetate	70	51	76	-	198	8,0	-	397
-propionate	1,6	1,7	3,9	-	4,3	n.d.	-	1,4
-lactate	2,2	35	35	-	62	n.d.	-	57
-butyrate	5,3	1,9	0,5	-	4,7	n.d.	-	5,3
-hexanoate	0,2	1,8	1,5	-	2,2	n.d.	-	2,3
-a-hydroxy-isohexanoate	n.d.	0,8	0,5	-	0,5	n.d.	-	1,2
-heptanoate	n.d.	0,02	n.d.	-	0,05	n.d.	-	0,05
-octanoate	1,0	2,4	1,5	-	7,7	0,4	-	10,3
-nonanoate	n.d.	0,09	0,05	-	0,07	+	-	0,3
-decanoate	1,3	3,4	2,5	-	9,9	0,6	-	14
-laurate	0,6	0,8	0,4	-	1,7	0,2	-	3,2
-myristate	0,1	0,1	0,04	-	0,1	0,06	-	0,4
-palmitate	0,1	0,2	0,08	-	0,1	0,06	-	2,5
-phenylacetate	n.d.	0,07	0,07	-	0,08	n.d.	-	0,1
-benzoate	0,02	0,09	0,03	-	0,04	n.d.	-	0,06
Diethyl -succinate	0,2	0,8	3,2	-	4,2	0,04	-	7,1
Methyl -acetate	1,2	1,5	0,8	-	7,8	n.d.	-	5,8
-hexanoate	n.d.	0,2	n.d.	-	0,03	n.d.	-	0,1
-heptanoate	n.d.	n.d.	+	-	0,04	n.d.	-	0,08
-octanoate	n.d.	0,04	0,04	-	0,1	n.d.	-	0,1
-decanoate	0,01	0,1	0,1	-	0,2	n.d.	-	0,3
-laurate	0,02	0,02	n.d.	-	n.d.	+	-	0,03
-myristate	n.d.	n.d.	n.d.	-	n.d.	n.d.	-	0,02
-palmitate	n.d.	n.d.	n.d.	-	n.d.	0,01	-	0,07
Propyl -acetate	0,1	0,2	0,2	-	0,3	n.d.	-	1,1
Butyl -acetate	0,8	3,3	0,7	-	3,7	0,1	-	2,4
Isobuty -acetate	2,1	0,3	0,2	-	2,1	n.d.	-	7,2
Isoaml -acetate	0,2	0,7	0,8	-	1,0	n.d.	-	1,0
-lactate	n.d.	0,4	0,4	-	0,7	n.d.	-	0,8
-hexanoate	n.d.	0,02	0,01	-	0,02	n.d.	-	n.d.
-octanoate	0,02	0,07	+	-	0,1	n.d.	-	0,6
-decanoate	+	0,03	+	-	0,06	n.d.	-	0,08
Hexyl-acetate	0,07	0,4	0,2	-	0,4	n.d.	-	0,2
2-Phenylethyl-acetate	+	0,1	0,2	-	0,5	n.d.	-	0,5
Benzyl-acetate	n.d.	0,02	0,01	-	0,01	n.d.	-	0,02

n.d. = none detected; + = <0,01 mg/100 ml ethanol

RASPBERRY BRANDY

Whereas in fruit and pomace brandies the alcohol derives from the fermented sugar of the fruits or pomace, in raspberry brandies the alcohol does not, or only to a minor part. derive from the berries, whose sugar content is very low. Raspberry brandy has to be produced from fresh or frozen and unfermented berries with the addition of alcohol, followed by distillation. In Germany neutral alcohol has to be used, but there is no regulation concerning the amount of raspberries per litre ethanol; in general the amount ranges from 100 g to 1500 g/l ethanol. In France the alcohol used has to originate from fruit; the use of alcohol originating from potatoes, grains or molasses is not permitted. The amount of raspberries has to be at least 100 kg per 12,5 l ethanol equaling 8000 g/l ethanol.

Responsible for the typical flavour of raspberries is 1-(4-hydroxyphenyl)-3-butanone, called raspberry ketone. Though the concentration of this real flavour impact compound in raspberries ranges from 0,4 to 2,4 mg/kg, it is not determinable in raspberry brandies, that is to say the concentrations are markedly lower than 0,01 mg/100 ml ethanol, even if high amounts of raspberries per litre ethanol are used. Distillation tests have shown, that the raspberry ketone practically does not pass over into the distillate in determinable amounts /40, 41/. Hence, it can be taken for granted, that raspberry brandies that contain raspberry ketone in concentrations of 0,01 mg/100 ml ethanol or more are adulterated by artificial flavouring.

Besides raspberry ketone some other components such as α- and ß-ionone are assumed to contribute to the raspberry odour as well /41, 3/. It was suggested that the concentrations of linalool, α-terpineol and cis-3-hexen-1-ol are suited to assess the amount of raspberries used per litre ethanol and to draw conclusions on the origin of the raspberries /42, 40/.

Quantitative data on the volatiles of raspberry brandies have been reported by several authors in the last few years /12, 42, 43, 40/. The concentrations of 53 components of French and German raspberry brandies are listed in Table 4 /43/. Of those components 24 could be indicated as originating from the original fruit, whereas the other components derived either from microbial metabolism when more or less fermented or bacterial infected berries were used, or from the alcohol when it was not neutral, or from enzymatic and chemical reactions during the production process. On average French samples contained much higher concentrations both of raspberry specific and non-raspberry specific compounds due to higher amounts of raspberries used per litre ethanol and the use of fruit alcohol with high levels of volatiles (Table 4).

Table 4 - Volatiles in raspberry brandies (mg/100 ml ethanol)

	German	French		German	French
Number of samples	18	7	Number of samples	18	7
	mean	mean		mean	mean
Methanol	67	225	Acetaldehyde	3,0	22
1-Propanol	2,6	28	Propionaldehyde	0,03	0,1
Allylalcohol	0,06	0,6	Isobutyraldehyde + Acetone	0,2	1,2
1-Butanol	**0,09**	**1,6**	**Hexanal**	**0,01**	**0,3**
2-Butanol	0,7	11	Acrolein	+	1,1
2-Methyl-1-propanol	2,7	44	**Acetoin**	**0,3**	**2**
Isoamylalcohols, sum	6,7	137	**Benzaldehyde**	**0,06**	**0,2**
1-Pentanol	**0,1**	**0,6**	Furfural	0,3	2,7
1-Hexanol	**0,1**	**1,7**	1,1-Diethoxyethane	1,9	11
cis-3-Hexen-1-ol	**0,2**	**0,6**	1,1,3-Triethoxypropane	0,01	0,1
2-Heptanol	**0,05**	**0,2**			
1-Octanol	**0,02**	**0,1**	Ethyl-formate + Methylacetate	0,9	3,1
1-Nonanol	**0,02**	**0,06**	-acetate	10	75
1-Decanol	+	0,04	-lactate	1,0	6,2
2-Phenylethanol	0,09	0,4	**-hexanoate**	**0,1**	**0,5**
Benzylalcohol	**0,2**	**0,2**	-octanoate	0,09	0,9
			-decanoate	0,3	1,2
Linalool	**0,2**	**0,6**	-laurate	0,3	0,6
Geraniol	**0,09**	**0,4**	-myristate	0,2	0,2
α -Terpineol	**0,07**	**0,3**	-palmitate	0,3	0,6
4 -Terpineol	**0,08**	**0,06**	**-benzoate +**	**0,1**	**0,7**
α -Ionol	**0,02**	**0,3**	Diethylsuccinate		
α -Ionone	**0,02**	**0,08**	**Butyl -acetate**	**0,03**	**0,3**
β -Ionone	+	**0,01**	Isoaml-acetate	+	0, 2
			Hexyl -acetate	+	**0,1**
γ -Hexalactone	**0,01**	**0,04**	**cis-3-Hexenyl-acetate**	**0,01**	**0,3**
Dimethylsulphide	0,12	0,26	2-Phenethyl-acetate	0,02	0,03
Alcohols,sum	81	451	Carbonyl compounds,sum	6	41
Terpenes,sum	0,5	1,8	Esters,sum	13	90
			Raspb. specific comp.,sum	1,9	11,3

Raspberry specific components are bold typed; n.d. = none detected; + = <0,01 mg/100 ml ethanol

CONCLUSION

Fruit brandies differ widely in quality and composition dependent on the raw material used and the processing procedures. They are especially rich in volatile components, both in number and quantity. Between the individual brandy groups there are considerable and characteristic differences as far as the occurrence and the concentrations of certain volatiles are concerned. The knowledge of the quantitative composition of fruit distillate flavours allows not only to characterize the different spirits, but also to assess their authenticity, purity and quality, to detect adulterations, and to derive better information regarding the condition and quality of the raw material used and the processing procedures applied.

REFERENCES

1. Nykänen, L.; Suomalainen, H. Aroma of beer, wine and distilled alcoholic beverages. Akademie-Verlag, Berlin 1983
2. Ter Heide, R. The flavour of distilled beverages. In: Food flavours Part B. The flavour of beverages (Morton, I.D.; MacLeod, A.J., Eds) Elsevier, Amsterdam 1986, pp. 239-336.

3. Dürr, P.; Tanner, H. Fruit flavours and their relevance to the flavour of the final distilled beverage. In: Flavour of distilled beverages (Piggott, J.R. Ed), Ellis Horwood Ltd, Chichester 1983, pp. 33-48.
4. Maarse, H.; Visscher, C.A. Volatile compounds in food. TNO-Division of Nutrition and Food Research, TNO-CIVO Food Analysis Institute, Zeist 1983-1986.
5. Postel, W. Volatile Components of fruit brandies. In: Flavour research of alcohole beverages, Proceedings Alko Symposium, (Nykänen, L.; Lektonen, P. Eds), Helsinki 1984, pp. 175-187.
6. Frank, D. Branntweinwirtschaft 1983, 123, 278-282.
7. Reinhard, C. Dt. Lebensm.Rdsch. 1978, 74, 299-301.
8. Tanner, H.; Brunner, H.R. Obstbrennerei heute. Heller-Verlag Schwäbisch-Hall 1982.
9. Pieper, H.J.; Bruchmann, E.E.; Kolb, E. Technologie der Obstbrennerei. Ulmer-Verlag, Stuttgart 1977.
10. Postel, W.; Adam, L. Alkohol-Ind. 1982, 95, 287-289, 304-306, 339-341, 362-363.
11. Velisek, J.; Pudil, F.; Davidek, J.; Kubelka, V. Z. Lebensm.Unters.Forsch. 1982, 174, 463-466.
12. Wencker, D.; Louis, M.; Laugel, P.; Hasselmann, M. Dt.Lebensm.Rdsch. 1981, 77, 237-238.
13. Bindler, F.; Laugel, P. Dt.Lebensm.Rdsch. 1985, 81, 350-356
14. Hildenbrand, K. Branntweinwirtschaft 1982, 122, 2-8.
15. Postel, W.; Adam, L. Branntweinwirtschaft 1983, 123, 414-420.
16. Postel, W. Dt. Lebensm.Rdsch. 1982, 78, 211-215.
17. Hieke, E.; Vollbrecht, D. Z.Lebensm.Untersuch.Forsch. 1980, 171, 38-40.
18. Tuttas, R.; Beye, F. Branntweinwirtschaft 1977, 117, 349-355.
19. Ismail, H.M.; Williams, A.A.; Tucknott, O.G. Z.Lebensm. Untersuch.Forsch. 1980, 171, 24-27.
20. Bricout, J. Ind. Aliment.Agric. 1977, 94,, 277-281.
21. Versini, G.; Inama, S. Vini Ital. 1983, 25, 43-50.
22. Postel, W.; Adam, L.. unpublished.
23. Postel, W.; Adam, L. Branntweinwirtschaft 1981, 121, 146-152.
24. Williams, L.A.; Knuttel, W.P. Computer modelling of aroma compound behaviour during batch distillation. In: Flavour of distilled beverages (Piggott, J.R. Ed), Ellis Horwood Ltd, Chichester 1983, pp. 134-144.
25. Postel, W.; Adam, L. Quantitative determination of volatiles in distilled alcoholic beverages. In: Topics in Flavour Reserach (Berger, R. et al.Eds). Eichhorn Verlag Marzling-Hangenham 1985, pp. 79-108.
26. Conacher, H.B.S.; Page, B.D. Ethyl carbamate in alcoholic beverages. A Canadian case history. In: Proceedings of Euro Food Tox II, Institute for Toxicology CH-8603 Schwerzenbach 1986, pp. 237-242.

27. Baumann, V.; Zimmerli, B. Mitt.Gebiete Lebensm.Hyg. 1986, 77, 327-332.
28. Christoph, N.; Schmitt, A.; Hildenbrand, K. Alkohol-Ind. 1986, 15, 347-354.
29. Bertrand, A.; Triquet-Pissard, R. Connaissance Vigne Vin 1986, 20, 131-136.
30. Christoph, N.; Schmitt, A.; Hildenbrand, K. Alkohol-Ind. 1987, 16, 369-373; 17, 404-411.
31. Andrey, D. Z.Lebensm.Unters.Forsch. 1987, 185,, 21-23.
32. Adam, L.; Postel, W. Branntweinwirtschaft 1987, 127, 66-68
33. Ough, C.S. J.Agric.Food Chem. 1976, 24, 323-331.
34. Baumann, V.; Zimmerli, B. Schweiz.Z. Obst- u.Weinbau 1986, 122, 602-607.
35. Mildau, G.; Preuss, W.; Frank, W.; Heering, W. Dt.Lebensm. Rdsch. 1987, 83, 69-74.
36. Tanner, H.; Brunner, H.R.; Bill, R. Schweiz.Z.Obst- u. Weinbau 1987, 123, 661-665.
37. Baumann, V.; Zimmerli, B. Mitt.Gebiete Lebensm.Hyg. 1987, 87, 317-324.
38. Ingledew, W.M.; Magnus, C.A.; Patterson, J.R. Am.J.Enol. Vitic. 1987, 38, 332-335.
39. Bertrand, A.; Barros, P. Comnnaissance Vigne Vin 1988, 22, 39-47.
40. Renner, R.; Hartmann, U. Lebensmittelchem.Gerichtl.Chem. 1985, 39, 30-32.
41. Tressl, R. Lebensmittelchem.Gerichtl.Chem. 1980, 34, 47-53.
42. Renner, R.; Zipfel, K. Lebensmittelchem.Gerichtl.Chem. 1981, 35,, 59-60.
43. Postel, W.; Adam, L. Dt.Lebensm.Rdsch. 1983, 79, 117-122.

12

The influence of the quantity of yeast in wine on the volatiles of grape wine brandies

W. Postel and L. Adam
Institut für Lebensmitteltechnologie und Analytische Chemie, Technische Universität München, D-8050 Freising-Weihenstephen, West Germany

Increasing amounts of yeast in wines lead to increasing concentrations of ethyl decanoate as well as ethyl octanoate and ethyl dodecanoate in wine distillates. To a considerable smaller degree this also holds true for ethyl hexanoate, ethyl myristate, ethyl palmitate, and for acetoin /1/ (Fig. 1).

The amount of yeast in wines has no or just a negligible influence on the concentrations of all other esters (including ethyl heptanoate, ethyl nonanoate, and the methylesters), of the carbonyl compounds, and of all alcohols and terpenes in wine distillates /1/ (Fig. 1).

The results are confirmed by analysis of an authentic yeast distillate.

REFERENCE

1. Postel, W.; Adam, L. Einfluß des Hefeanteils in Wein auf die Gehalte an flüchtigen Verbindungen in Weindestillaten. Dt.Lebensm.Rdsch. 1984, 80, 267-273

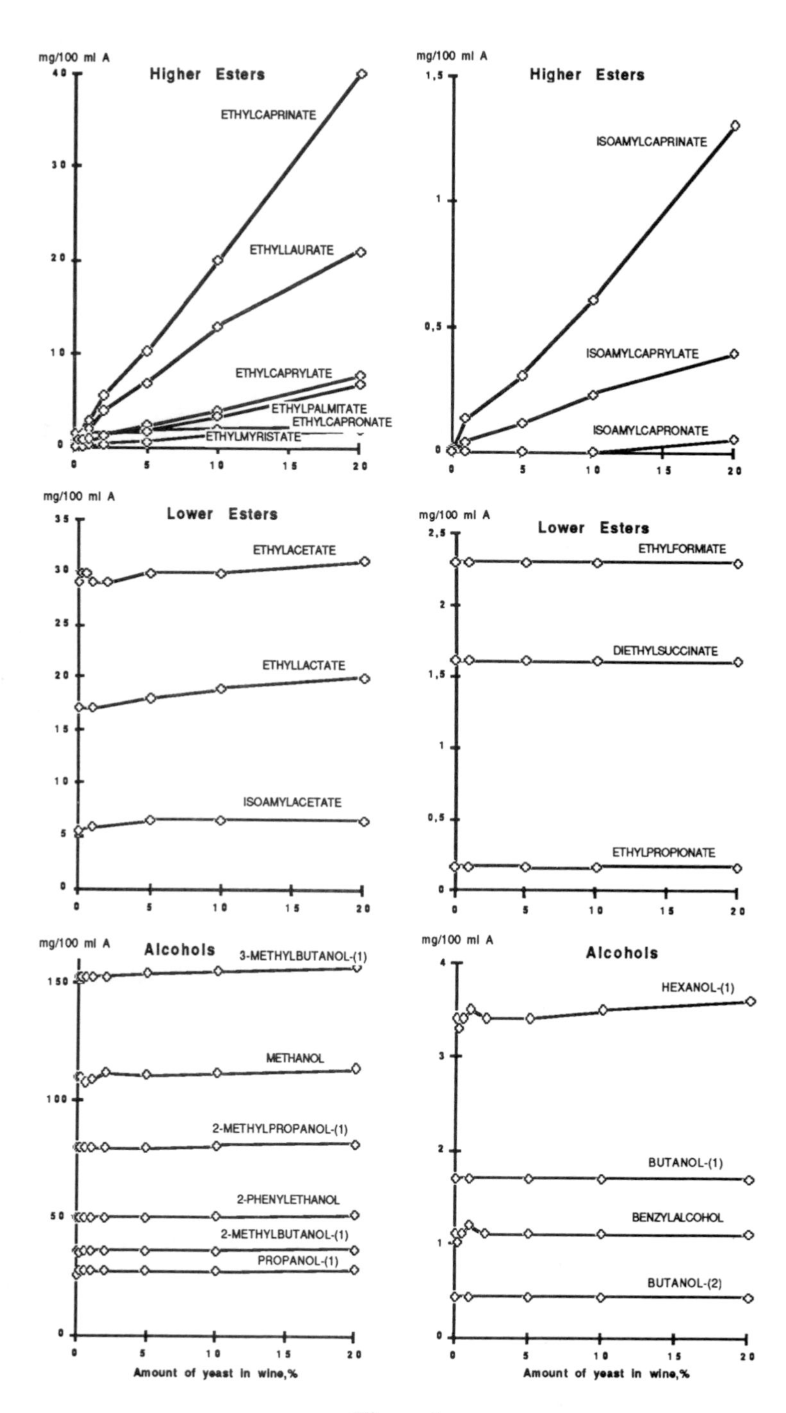
mg/100 ml A
Higher Esters
ETHYLCAPRINATE
ETHYLLAURATE
ETHYLCAPRYLATE
ETHYLPALMITATE
ETHYLCAPRONATE
ETHYLMYRISTATE
ISOAMYLCAPRINATE
ISOAMYLCAPRYLATE
ISOAMYLCAPRONATE
Lower Esters
ETHYLACETATE
ETHYLLACTATE
ISOAMYLACETATE
ETHYLFORMIATE
DIETHYLSUCCINATE
ETHYLPROPIONATE
Alcohols
3-METHYLBUTANOL-(1)
METHANOL
2-METHYLPROPANOL-(1)
2-PHENYLETHANOL
2-METHYLBUTANOL-(1)
PROPANOL-(1)
HEXANOL-(1)
BUTANOL-(1)
BENZYLALCOHOL
BUTANOL-(2)
Amount of yeast in wine,%

Figure 1

13

The contributions of the process to flavour in scotch malt whisky

Alistair Paterson and John R. Piggott
Food Science Division, Department of Bioscience and Biotechnology, University of Strathclyde, 131 Albion Street, Glasgow, G1 1SD, Scotland

1. INTRODUCTION

As a distilled beverage, whisky has three predominant influences on the flavour of the retailed product -

1. The raw materials and the details of the manufacturing process.

2. The casks that are used for maturing the distilled spirit, particularly their origin and treatments prior to this filling.

3. The view of the distiller or blender or his marketing department on what his product should be.

Since the third of these factors is of over-riding importance in determining the flavour of blended whiskies, which consist of malt and grain whiskies in varying proportions, the emphasis of this paper will be upon malt whiskies. These, in addition to their own market share, act as the major flavour determinants in the blended whiskies that form the bulk of whisky production. In this paper we will review significant causes of variation in flavour between distilleries and between batches of distillate, and the current state of our knowledge on what influences these.

Assessing malt whiskies is traditionally a process in which the distiller uses his nose as an instrument to analyse samples from a number of individual casks before deciding upon a mix that will generate a batch of his whisky. This blender's experience, viewpoint and prejudice, linked with the palette of individual cask flavours available, will generate the final commercial product. In this way a single malt differs from the single-cask whiskies now available from specialist sources.

The Scotch malt whisky manufacturing process

Stage	Input and Variation
Grain	Barley - variety and seasonal
Malting	Water - area and seasonal
Kilning	(Peat smoke) gas contaminants
Mashing	Water - area and seasonal
Fermentation	Yeast - strains, contaminants
Distillation	Feints - operator practice
Maturation	Casks - availability
Blending	Opinion - available batches
Bottling	Bottle material - marketing

Figure 1. Flow diagram of malt whisky production.

Despite the universality of the basic malt whisky manufacturing process, shown in Figure 1, considerable variation is observed in the product. Thus differences in practice generate a range of beverages varying diversely in character.

In our laboratory, assessments of whisky flavour are made using descriptive sensory analysis. A selected vocabulary of adjectives describing individual aroma notes is utilised by a panel of trained and experienced assessors to describe a seof samples and produce a 3-way matrix of data. These data are processed using appropriate statistical procedures to generate profiles of individual samples that can be presented in varying ways [1].

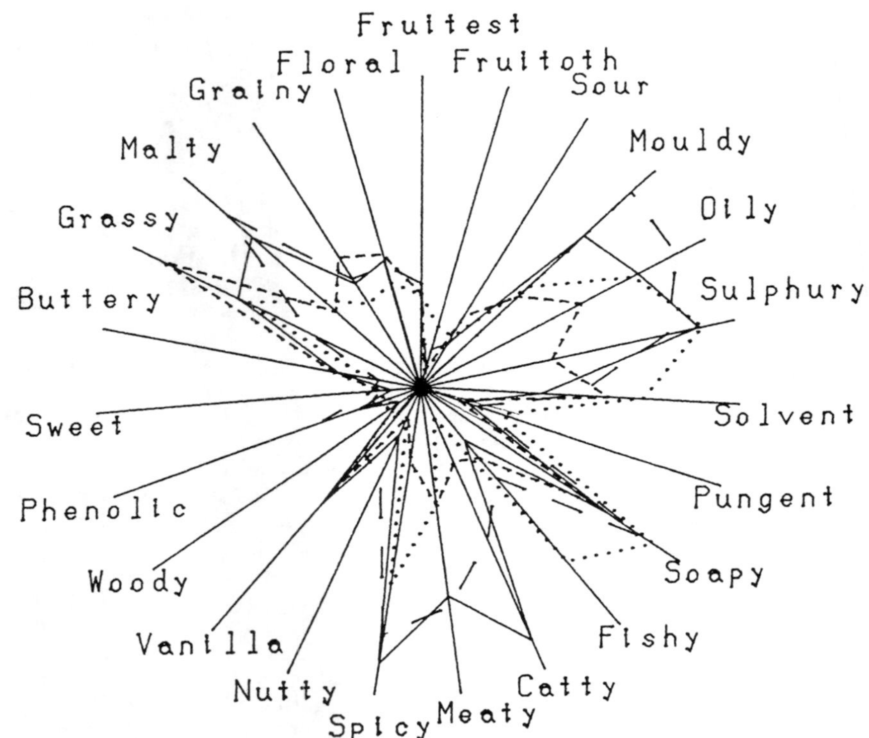

Figure 2. Profiles of four whisky distillates shown as 'spider's web' plots.

Whisky vocabulary

Pungent	Sour
Solvent	Nutty
Spicy	Buttery
Grainy	Grassy
Malty	Phenolic
Mouldy	Oily
Fruity (estery)	Woody
Fruity (other)	Meaty
Floral	Sulphury
Smooth	Catty
Vanilla	Fishy
Soapy	Sweet

Our current vocabulary, developed for over a decade, is presented in Figure 2. Utilising this, product profiles can be presented for any whisky in relation to its competitors or stablemates and profiles of a range of whiskies are shown in one form of presentation.

Since each whisky is analysed by nosing a sample, diluted to 23% alcohol with water, held in a standard glass [2], each whisky flavour profile is a summation and normalisation of the observers' detections of aroma compounds present in the beverage headspace. The contribution of each aroma compound to the whole will vary according to the concentration of the compound in the headspace, its detection threshold, its intensity function, and its ability to interact with other aroma compounds at the levels in which they are present, both above and below their threshold level for individual detection. It has been generally considered that the major contributors to spirit quality and present at ppm level are the alcohols, esters, fatty acids and phenols. However, preparation of solutions of these substances at the levels reported in Scotch whisky yield preparations with distinctly different sensory characteristics from those of whisky. Furthermore, when the odour intensity of the solutions or of the component substances is compared utilising threshold measurements it is found that the greater part of the whisky aroma cannot be explained. An alternative viewpoint that is gaining acceptance is that variations in character and quality of both malt and grain spirits is largely dependent on the presence of substances at concentrations of 0.1 ppm or less and in grain whisky at below 0.1 ppb. Some of these have odour intensities many orders of magnitude greater than the major components conventionally detected by gas chromatography. Consequently the distiller has traditionally and currently carries out sensory evaluation to examine the nature and concentrations of aroma compounds built up in the glass headspace.

2. THE MANUFACTURING PROCESS: SOURCES OF VARIATION

2.1 Barley, malting and kilning

Since 1678, when Sir Robert Moray described in an article the current state of knowledge of factors affecting malting and malt quality on the basis of practical experience, both malting practice and barley properties have been changing [3]. Even since the last war, constant change has been taking place in malting. Currently, the predominant process in Scotland is the "Saladin Box" in which large concrete boxes and circular steel boxes, holding 200 - 2000 tons of barley, are used to soak and germinate barley. After an appropriate period the seedling is dried and killed by a hot air stream in a gas-fired kiln and during this stage important malt aroma and flavour compounds are developed.

The final malt can be regarded as a barley product consisting of starch; proteins and amino acids; glucans, pentosans and related polymers; lipids; cell wall-derived phenolics; and malt and smoke flavour compounds. Each of these can contribute to defining the character of the yeast fermentation and thus the final distillate flavour and each will occur at levels determined by the grain and the detailed procedure used by the maltster.

Agronomy and breeding programmes have produced considerable changes in the components of the barley used in malting yet much of grain composition are, in practice, ignored. Typically, a maltster will choose a barley on the basis of fermentable extract yield, protein content, corn size, and germination capacity based on micromalting performance. Hot-water or fermentable extract yield is particularly important since the spirit yield of barley malt (litres alcohol/tonne) is calculated by multiplying the hot water percentage extract (typically 74 - 79% at a moisture content of 4.5 - 5.0%) by the figure of 5.22 [3]. Barleys for malting are also chosen as having a protein content of 1.6% (+/- 0.1%) and for having grains larger than 2.2mm in size. However, the varieties used by the maltster may change over short periods on the basis of improvements in yield, new requirements with respect to disease resistance, and the balance between spring and winter crops amongst other factors and each new barley variety appears to change the flavour of the whisky obtained, albeit in a subtle way. Thus many distillers specify the variety of barley to be used in production of malt for their whisky, and ignore batch variation in flavour progenitors.

Kilning is the major stage of the process where flavour compound production and destruction are observed. Reactions considered to be key are the enzymic and chemical oxidation of unsaturated fatty acids, the combination of free amino acids and reducing sugars in Maillard reactions and breakdown of thermo-labile sulphur compounds formed in protein biosynthesis [4]. Many of the resulting organic nitrogen and sulphur compounds have very low thresholds for detection and high intensity functions for the human nose and through their volatility are carried into the spirit.

The major part of our current understanding of the formation of kiln-derived flavours is based upon study of brewing malts [4,5], particularly in Germany where the subtleties of beer flavour are of widespread interest. Yet it is clear that the contribution of the malt to whisky flavour must differ since brewing worts are boiled together with hops before pitching of yeast. The malt flavour compounds in whisky worts are merely cooled with the distillation of the wash acting as a selection. However, it cannot be ignored that peat-smoked malts yield distinctive whiskies and mould in the malt can be detected after a prolonged maturation of the spirit suggesting that the flavour contribution of malt can be underestimated.

One of the predominant fatty acids in barley lipids is linolenic acid (18:2) which during malting is modified by lipoxygenase action to form 9- and 13-hydroperoxides that are further converted to aldehydes such as hexanal and trans-2-hexenal. Sugars and amino acids obtained from amylase, glucanase and protease action combine and following Strecker degradations are observed as unsaturated aldehydes, furans, pyrroles and sulphur compounds such as the sensorally important di- and tri-sulphides in the final malt. Levels of the precursors of all of these compounds are strongly influenced by varietal character and environmental factors that influence the proportions of the classes of storage proteins of the barley, each with a distinctive balance of amino acids, containing nitrogen and sulphur contributing to aroma compound production.

A further factor in generating malt variability is the optional introduction into the kiln air of smoke compounds, produced by destructive pyrolysis of peat humic and fulvic acids derived from complexation of lignocellulose with proteins during the humification process. In earlier times, the distillery staff would spend a part of their working year digging the peat and drying it in preparation for the kilning in the seasonal malting. Currently, the predominant indirect heating of the kiln air by heat exchangers and utilisation of gas or oil as energy source, requires peak smoke to be added to the air stream. However, because peat contains protein-derived material from the original plant, nitrosamines are formed in the gas phase during peat pyrolysis. To reduce the concentration of these carcinogenic compounds, rock sulphur is burned simultaneously with the peat or gaseous sulphur dioxide added to the gas stream, so that 10 - 30 ppm of sulphur dioxide is detected in the final malt. "Peating" in itself requires skill on the part of the maltsters since the smoke is adsorbed most effectively when the malt has a moisture content of between 15 and 30% and is dependent upon the intensity of smoke in the air stream.

Traditionally malts are defined with respect to peating by total phenol content using the following categories [3]:

lightly peated	1.0 - 5.0	ppm total phenols
medium peated	5.0 - 15.0	ppm total phenols
heavily peated	15.0 - 50.0	ppm total phenols.

Obviously phenols differ in their contribution to flavour according to their detection thresholds and intensity functions and peats also differ in composition through varying origin. Consequently, the smoke contribution to malt flavour may be one of the most significant in determining the character of the beverage. It has been calculated that only about 22% of the malt volatile phenols appear in the malt distillate, most phenols (including phenol) being water-soluble remaining in the residue after distillation. However many of the other organoleptically significant classes of compounds present in peat smoke, notably the pyridines and thiazoles, are recovered with greater efficiency and have lower threshold concentrations and may contribute more to the final flavour than the phenols *per se*.

2.2 Mashing

The production of whisky requires a clear wort to prevent suspended material "burning" at still surfaces later in the process. Consequently, mashing is important in that the maximum of soluble fermentable carbohydrate that can be converted to alcohol must be extracted from the malt with the minimum of insoluble material. Currently, the traditional deep batch mashtun and shallower lauter tun, used by brewers, are used and the malt itself is the primary filtrant [6]. Consideration of the detailed geometry of the mash vessel and degree of agitation are important in achieving high fermentable extract yield and in determining the extent of extraction of non-carbohydrate malt components since the malt itself is utilised as the filtrant.

A typical programme for mashing in malt whisky production would be as follows [6] :

1. 5 - 11 tonnes of malt are milled and loaded into the tun.
2. First water : 4-4.5 tonnes water/tonne malt is added at *ca.* 64°
3. Drainage so that 80 - 85% of the water is removed.
4. Second water : 1.5 - 2 tonnes water/tonne malt is added at 72 - 74°
5. Drainage.
6. Third and fourth waters are added at 80 - 90° with the combined volume exceeding that of the first water.

The total cycle time, including filling, mixing, draining and solids discharge, is 8 - 11 hours.

Lauter tuns differ from the conventional tun in having shallower beds and differences in the design of the rakes that cut, loosen and turn the beds of solids. Such mechanical differences will tend to generate differential extraction of malt components and in particular yeast volatile ester synthesis during fermentation is known to be strongly dependent upon the mashing protocol followed [7].

Obviously the distiller's first interest is to obtain the theoretical yield of fermentable material from his malt. A second major interest is the maintenance of good hygiene since significant levels of yeast and bacterial contaminants carried over from malt and tun in the unboiled whisky wort will contribute to the fermentation, reducing alcohol yield and generating products altering the flavour of the resulting spirit. A further desirable property is that the "secondary conversion" of enzymic depolymerisation of carbohydrate to fermentable sugars by the malt enzymes continues during and after mashing, in the fermentation. The pH of the wort is critical in this hydrolysis since acid conditions will inhibit the action of the hydrolytic enzymes. In grain whiskies, where thermotolerant bacterial contamination can result in acid production in the fermentation, pH variation can be significant.

2.3 Fermentation

The fermentation of the aqueous malt extract has been divided into three phases [8] :

1. Active yeast growth and acceleration fermentation phase.

2. Linear fermentation phase with limited growth.

3. Stationary or decline phase.

Fermentation has often been regarded as the central stage of flavour compound generation in whiskies since it is the yeast action that leads to formation of such compounds as higher alcohols, fatty acids and esters which appear at relatively high concentrations in the final beverage [9]. The amount of yeast added is critical in flavour compound development, with the optimum for a malt fermentation being several times that necessary for that in grain whisky production. Alcohol production and yeast and bacterial growth all occur to an extent in competition and the kinetics of each will affect the range and concentration of compounds entering the still.

Distillers' yeasts (all now classified as Saccharomyces cerevisiae) are very versatile microorganisms, being capable of fermenting at temperatures of up to 46°C in the pH range 3 - 10, in the presence of 0 - 15% alcohol and 0.1 - 25 % sugar. In strain development, efficient utilisation of carbohydrate, rapid fermentation, high ethanol yield and tolerance, good product flavour and resistance to other microorganism are major selection factors. However, the detailed metabolism of the organism will vary according to the growth substrates available, the temperature and pH of the wort, and the growth phase. Flavour assessments can only be made by sensory evaluation of the new distillate which will not directly coincide with flavour in the matured beverage.

It is considered desirable that a mixture of yeasts should be used in to ensure a good beverage character [9]. The dominant primary yeast, that concerned primarily with ethanol production, is the Distillers (Yeast) Company "M" strain. Secondary yeast strains which are lower in yield, slower in fermentation, and less ethanol tolerant are added to worts to improve flavour and for reasons of economy. Distillers' yeasts alone are added typically at 0.35% (w/v), whereas the mixed primary and secondary yeasts are added at a ratio of about 0.42%. Often spent brewers' or bakers' yeasts are used in the secondary role since they produce more flavour compounds than distillers' yeasts and are cheaper than specialised secondary yeasts. Secondary yeasts also differ significantly metabolically from the primary yeasts since the brewing yeasts will have been obtained by anaerobic propagation in the beer fermentation whereas the primary yeasts will be raised by growth in aerated fermenters. Bakers' yeasts differ again in that they are produced commercially with a terminal highly-aerated growth phase with limiting carbon source, yielding metabolically fully oxidative cells.

The timing of the phases of yeast metabolism in the whisky fermentation are strongly dependent upon the character of the wort. The primary growth phase, biomass production, is strongly dependent in length on the availability of utilisable nitrogen for protein biosynthesis. This in turn is dependent upon protease activity in the original malt. The major soluble endproducts of a whisky fermentation have been estimated as approximately :

Ethanol	92.1%
Glycerol	5.8%
Succinate	0.9%
Yeast (dry weight)	1.2%

The production of the flavour compounds which form a minor part by weight of the fermentation products been summarised as being predominantly through synthesis of [10] :

1. higher alcohols
2. fatty acids
3. esters
4. carbonyl compounds

1. **Higher alcohols** are formed from catabolism of glucose and amino acids in the wort. Amyl alcohol, isobutanol, propanol and 2-phenylethanol are formed initially during the exponential phase and continue to be produced in the linear phase. Individual strains appear to differ in the levels of individual higher alcohol production whereas changing the grist is thought to influence the level of total higher alcohol production but not the ratio of the individual components [11]. Factors that increase the rate of fermentation, stimulate production of higher alcohols [12].

2. **Fatty acids** are also secreted at significant levels during fermentation with the proportions of the individual acids being typical observed in the fermented wash as shown in Table 1.

Acetic acid	50 - 90% Total acids
	% Total acids excluding acetic.
Propionic acid	1.5%
Isobutyric acid	4.9%
Butyric acid	1.5%
Isovaleric acid	5.9%
Hexanoic acid	4.2%
Octanoic acid	26.7%
Decanoic acid	31.6%
Dodecanoic acid	16.2%
Tetradecanoic acid	2.2%
9-hexadecenoic acids	2.1%

Table 2. Fatty acids derived from fermentation.

The origins of the differing acids appears to vary with long chain fatty acids being synthesised from acetyl coA, malonyl coA and NADPH. Medium chain fatty acids appear to enter the wash by leakage from cells of intermediates of long-chain fatty acid biosynthesis. Of the short chain fatty acids some, such as acetic acid, are derived carbohydrate metabolism whilst others obtained by reactions such as decarboxylation of keto acids e.g. butyric and _iso_valeric acids [12].

The even chain number fatty acids are detected in fermented worts as "caprylic" flavours in beer or odours of musty, soapy, fatty, oily, and rancid with typical threshold concentrations for detection in ppm: 4.5ppm for octanoic acid, 1.5 ppm decanoic acid and 0.6 ppm dodecanic acid. However, in flavour terms, the contributions of these compounds are additive and it is known that a significant component of commercial beer flavour is this "caprylic" note [13].

3. _Esters_ are flavour compounds thought to arise from enzyme-catalysed reactions between acetyl CoA and the free alcohols and give rise to "fruity" notes in whiskies. Acids and esters, however, also contribute to ester formation in a ratio related to concentrations of the acids available for reaction. The concentration of free acids can vary considerably through differences in mashing procedure since variation in raking in the mash tun influences the release of a fatty-acid rich fraction from the malt [14]. Addition of exogenous oleic acid and increased oxygen content are known to enhance formation of medium chain fatty acids and ethyl esters although ester synthesis appears to be very strain dependent with individual strains producing very low levels being observed.

4. _Carbonyls_ are known to be generated in the kilning of malt, typically through breakdown of amino acids as methional, a Strecker degradation product of methionine. Other important malt

carbonyls derived from amino acids are 3-methylbutanal, 2-methylbutanal, 2-methyl propanal (primary malty odour) and 2-phenylethanal. These are reduced by the yeast during fermentation into flavour-active carbonyls which appear in the final spirit.

Compound	Concentration range
Isobutanal	0.06 - 0.64 ppm
Isopentanal	1.2 - 12.8 ppm
2-Methylthioethanal	0 - 0.04 ppm
2-Phenyl-2-butenal	0.03 - 0.3 ppm
4-Methyl-2-phenyl-2-pentenal	0 - 0.02 ppm
4-Methyl-2-phenyl-2-hexenal	0 - 0.01 ppm
5-Methyl-2-phenyl-2-hexenal	0 - 0.06 ppm
2-Furfural	0.45 - 2.7 ppm
2-Acetylfuran	0 - 0.54 ppm
2-Formylpyrrole	0.05 - 0.58 ppm
2-Acetylpyrrole	0.02 - 1.40 ppm
2-Acetyl-5-methylpyrrole	0 - 0.02 ppm
2-Formyl-5-methylpyrrole	0.02 - 0.40 ppm
2-Acetylthiazole	0.06 - 1.2 ppm
2-Acetyl pyridine	0.01 - 0.02 ppm

Table 2. Strecker aldehydes and Maillard reaction products in beer malts that are reduced by yeasts in fermentations [15].

'Diacetyl' a sweetish odour is associated with yeast strain character and pH amongst other fermentation conditions. It is formed from acetolactate which is an intermediate in valine biosynthesis in yeasts and is formed particularly in very rapid fermentations. In beer it is known to give a buttery aroma at 0.15 ppm [16].

4. Phenols are formed from coumaric and ferulic acids of the barley cell-wall either by decarboxylation by the yeast or through the boiling of the wash in the pot still. 4-Vinylguaiacol is a typical product obtained by decarboxylation of ferulic acid and is known to give a spicy, clove-like flavour to certain Bavarian beers [17].

Bacterial flavours are also produced during the fermentation. It has been calculated that a fully fermented 30-hour wash will contain upwards of 100 million bacteria ml^{-1} [9]. Some of these bacteria will reduce spirit yield, producing acids in the wash. Such bacterial activity can be monitored using pH monitoring or lactate specific electrodes. Many of these acids, however, are removed as precipitates at the chill-filter stage in bottling.

'Acrolein' unpleasant acrid notes are associated with bacterial infections of the wash and tend to be associated with long fermentations in which lactic acid bacteria build up. The bacteria produce a flavour precursor, hydroxypropionaldehyde, that is converted in the still to acrolein. A sour or sickly odour from n-butyric acid is sometimes formed by bacterial action if mashing temperatures are too low or if the wash is allowed to stand in cast iron vessels.

2.4 Influence of distilling practice on flavour

The traditional pot distillation is a two-stage process with each pear or onion shaped copper still containing 25,000 - 50,000 litres. The first (wash) still boiling the fermented wort for 5 - 6 hours to produce the first (low wines) distillate which concentrates the alcohol three-fold yielding a 26% alcohol spirit. The second (spirit) still produces three fractions :

1. The early fraction or foreshots which are recycled.
2. The Whisky at approximately 72 - 75% alcohol.
3. The feints which are distilled over until the alcohol concentration is reduced to 0.10 - 0.15% and recycled with discard of the residue.

In this second distillation, the timing of the taking of the three fractions is critical in determining the flavour of the new distillate. The esters appear in the vapour early in the distillation and decrease with time, whereas the higher alcohols appear later in the distillation. The timing of each cut is generally judged by the operator on the basis of hydrometer readings and spirit colour. The foreshots, in particular, are critical because they contain the most volatile and undesirable compounds from the distillation and the high molecular weight, least volatile compounds retained in the condenser or pipework from the feints of the previous batch. The only test commonly used is the mixing of equal volumes of distillate and water with the absence of cloudiness being regarded as determining the cut point. The cut to feints is determined by the alcohol content required in the spirit and prolonging the distillation will not only reduce the alcohol content of the new distillate but increase the concentration of higher alcohols.

In chemical terms the distillation is important in that selection of flavour compounds takes place by virtue of the fractionation of the still and sensitive organoleptically important molecules will be broken down or react with other components during the heating process in the wash, in the gas phase and on the copper surface of the still, the worm or the pipework.

'Burnt' and 'gassy' notes in new distillates are caused by organic sulphur compounds with very low odour thresholds. They are particularly important in new stills during commisisning and are known to be to an extent broken-down by catalytic action of the copper of the still [9].

2.5 Maturation chemistry

The accumulated flavour compounds in the final malt whisky beverage number some hundreds although many of these are now thought to be of limited importance. Yet it cannot be ignored that compounds even at concentrations below their threshold for individual detection, act alone and in combination to give flavour notes.

The most significant individual contributions to flavour in the new distillate are currently considered to be from:

1. Esters.
2. Lactones.
3. Carbonyls, notably long-chain unsaturated aliphatic [18]
4. Sulphur-containing compounds [19]
5. Nitrogen-containing compounds [20]
6. Phenols, derived from malt and from smoke.

Consequently in the maturation chemistry it is the behaviour of these compounds, their interaction with components eluted from the wood, and their interaction with the inner surface of the cask that is of critical importance in dictating the final product.

Scotch whisky casks vary considerably. Four principal categories have been defined:

1. The 558-litre American oak puncheon, charred on the inner surface in the U.K. and used first for grain spirit.

2. The first-fill wine-treated dumphogshead (254 litres). These were made using staves from American bourbon casks.

3. American standard barrel (191 litres) normally increased to 254 litres by addition of extra staves and new American oak ends and wine-treated before use.

4. The 500 litre American or Spanish Oak Butt.

Consequently both American (*Quercus alba*) and Spanish (*Q. sessilis*) as well as French (*Q. robur*) are used, each with varying histories. Differences between oaks are known to be critical in developing flavour in other beverages. In comparisons of brandy aged in French limousin oak or new American oak casks the presence of "oak lactone" methyl-octalactone, was not observed in spirit matured in French oaks but was important in those from American oaks [21]. Oak lactones are also observed in mature whiskies and appear to be important because they have low odour thresholds - 500 ppb and 70 ppb in 23% ethanol for respectively, *cis* and *trans* isomers, imparting "woody" notes to whiskies [22].

Whisky casks are traditionally treated to thermally-degrade the internal surface of the cask. Firing or mild charring are applied to casks from Spain whereas American Bourbon staves are heavily charred. There are three objectives in this thermal degradation of the wood.

First, a layer of active carbon is produced which can remove flavour elements, regarded as "undesirable", during the maturation. Secondly, oak lignin is anaerobically degraded in the layer immediately under the charcoal and flavour compounds such as vanillin are released and to dissolve in the spirit during the maturation. Finally, the total wood extract gives colour and phenols to the maturing whisky and through oxidative interactive reactions, new flavour components are produced.

The changes in that take place in the oak cask have recently been summarised by Nishimura [23] as :

a. Direct extraction of wood components.

b. Decomposition of wood macromolecules from the woody cell wall - lignin, cellulose and hemicellulose followed by elution.

3. Reactions of wood components with spirit components.

4. Reactions involving only the extracted wood materials.

5. Reactions involving only the distillate components.

6. Evaporation of low-boiling compounds through cask wood.

7. Physical rearrangement of components of the spirits to give structured liquid crystals.

Wood components are thus of critical importance to the final beverage. The heartwood of American white oak (Q.alba) is composed (cellulose (49 - 52%), lignin (31-33%), hemicellulose (approx. 22%) and a fraction extractable by hot water or ether (7-11%) which contains volatile oils, acids (some volatile), sugars, steroids, tannic substances, pigments and inorganics.

The water-soluble polyphenols that can be subdivided into:

i. Condensed tannins, derivitives of flavanols.
ii. Hydrolysable tannins that yield gallic and ellagic acids.

The volatile fraction has been shown to contain more than 100 components of the volatiles have been identified to date [24,25] and these are summarised in Table 3. The major proportion of volatile oils by weight are the oak lactones, 4-nonanolide and eugenol.

35 aliphatic compounds	:	7 hydrocarbons
		31 acids
		4 others
54 aromatic compounds	:	28 hydrocarbons
		17 phenols
		2 alcohols
		3 acids
		4 other compounds
3 furan compounds		
26 terpene compounds		

Table 3. Volatile components of maturation woods.

Lignins are present in cell walls as a complex with the hemicellulose, forming a resin that surrounds the cellulose fibres which give the cell wall its structural strength. This insoluble heteropolymer, glycolignin, has bonds that are hydrolysed with prolonged exposure to the spirit alcohol yielding soluble lower molecular weight fractions referred to as Braun's native lignins. These elute into the beverage to yield monomeric cinnamyl and methoxylated phenolic alcohols which are further oxidised to the sensorally important cinnamic and phenolic aldehydes. A summary of the pathways thought to be operating has been given by Reazin [26].

Consequently it can be seen that both the choice of wood and the barrel treatment can influence the flavour contribution of the barrel and this is widely seen in practice since over-used and maltreated casks do not produce whiskies of acceptable quality. However, it is currently widely accepted that the newdistillate has the most profound effect on the concentration of volatile flavour components in the final beverage. Duncan and Philp have claimed that few of these changed during maturation experiments [27] except for acetal and ethyl acetate although as predicted colour, solids, acids, esters and sugars in the spirit were extracted from barrels, in greater concentrations from those wine-treated in the cooperage. Yet little has appeared in the published literature on the sensory changes effected on spirits during maturation under varying conditions and with differing new distillates.

Cask and warehouse effects have recently been summarised as altering the following factors in malt whiskies with rates of maturation inversely related to spirit strength: pH, colour, solids content, total acids, esters, aldehydes, phenols, tannins and sugars [28].

3. CONCLUSIONS

In the past decade, considerable advances have been made in our understanding of the chemical composition of whiskies, largely through developments in analytical methodology that have made possible the detection of compounds at levels of parts per billion. However, what this has provided is a large amount of data without the means to relate the concentrations of trace chemicals to sensory data, that perceived by the blender or consumer.

Most GC chromatograms show data obtained from a selective detector, the flame ionisation detector (FID), although increasingly total ion chromatograms, obtained from mass spectrometers linked to gas chromatograms (GC/MS) are collected. The nose is also a selective detector, either when used to assess the quality of a whisky sample, or when linked to a capillary gas chromatograph by an olfactory outlet, with different responses from the FID. On the other hand, the mouth is arguably the primary detector used by the consumer.

It has been shown that approximately one third of the total information in sensory analyses of whisky, utilising trained panellists, came from the flavour contribution obtained by mouth[2]. This data is normally not collected because of the practical problems of organising experiments in which observers consume alcoholic beverages during working hours. Yet the consumer left to his own sensory profiling with his whisky glass tends to generate such descriptors as smooth, mellow, velvety, rounded, mild, rich and soft, as shown by members of our research group elsewhere in this volume (Consumer Free-Choice Profiling of Whisky). Such terms are considerably more difficult to equate with concentrations of individual spirit components than aroma notes and it has been suggested that the absence of off-notes and the mouthfeel and the retronasal component of beverage quality are what the average drinker primarily perceives. Thus an important conclusion must be that relatively little of what influences perception of whisky quality is currently understood.

The contribution of the malt and the maturation probably stand out from the other components of the whisky process in sensory terms. Any observer can distinguish an Island whisky with a heavily- peated malt from a Lowland whisky with an unpeated malt. Similarly there are few problems in distinguishing a whisky that has not been matured, by a defective or exhausted cask, even by the casual drinker. The sense of taste is thought to be relatively complex, with contributions from the tongue, the skin of the mouth and the nasal mucosa [29]. The perception of flavour in a beverage such as a wine or spirit will have a temporal sequence of flavour notes on different parts of the sensory surface. A part of this must be related to perception of topnotes and backnotes, as can be explored by considering the flavour of a sample of whisky over a period of an hour. This division of sensory perception into flavour and mouthfeel contributions from compounds remains to be explored, but volatile whisky components such as esters and alcohols must make a significant contribution to these facets.

Piggott and Findlay [17] have shown that three groups of interactions between flavour compounds can be observed when studying sensory data obtained from analysis of mixtures of esters found in whisky: synergism, with thresholds for mixtures being lower than individual detection thresholds; suppression at all mixing levels; and compounds that show either synergism or suppression depending upon their relative concentration.

The linking of instrumental and sensory analyses is a challenge for the whisky researcher. Modern gas and liquid chromatographs are able to detect trace spirit components, detectable previously only by the nose. Sensory analysis remains to be developed to allow the study of the contribution of these components to the beverage flavour. In particular the contribution of the malt flavour compounds and components such as Maillard reaction products including Strecker degradation products to whisky flavour remains to be explored. It has been calculated that two thirds of all known food flavour compounds are Maillard reaction products or product derivitives. The nitrogen and sulphur-heterocyclic compounds contribute important notes [19,20] which will change with the maturation of the new distillate to the beverage. This maturation process, since it generates mouthfeel and complex sensory perceptions readily discernable to the consumer, requires further elucidation.

4. ACKNOWLEDGEMENTS

The authors are grateful for support from Chivas Brothers Ltd and the Agricultural and Food Research Council for support.

5. REFERENCES

1. Piggott, J R (1986) (editor)
   Statistical Procedures in Food Research. pp. 416.
   London: Elsevier Applied Science.
2. Piggott, J R & Jardine, S P (1979)
   Descriptive sensory analysis of whisky flavour.
   J Inst Brew. **85**, 82 - 85.
3. Bathgate, G N & Cook, R (1989)
   Malting of barley for Scotch whiskies.
   In: The Science & Technology of Whiskies.
   (Eds. Piggott, J R, Sharp, R. & Duncan, R E B)
   London: Longman. In the press.
4. Tressl, R, Bahri, D & Helak, B (1983)
   Flavours of malt and other cereals. pp. 9 - 32
   In : Flavour of Distilled Beverages - Origin and Development.
   (Ed. Piggott, J R). Chichester: Ellis Horwood.
5. Tressl, R, Helak, B, Martin, N & Rewicki, D (1985)
   Formation of flavor compounds from L-proline. pp. 140 - 159
   In : Topics in Flavour Research (Ed. Berger, R G, Nitz, S, & Schreier, P). Marzling-Hangenham: H Eichhorn.

6. Wilkin, G D (1989)
Milling, cooking and mashing.
In : The Science & Technology of Whiskies.
(Eds. Piggott, J R, Sharp, R & Duncan, R E B).
London: Longman. In the press.
7. Berry, D R & Watson D C (1987)
Production of organoleptic compounds.
In : Yeast Biotechnology.
(Eds. Berry, D R, Russell, I & Stewart, G G).
London: George Allen & Unwin.
8. Kirsop, B H & Brown, M L (1972)
Some effects of wort composition on the rate and extent of fermentation by brewing yeasts. J Inst Brew **78,** 51 - 57.
9. Hardy, P J & Brown, J H (1989)
Process control. In: The Science & Technology of Whiskies.
(Eds. Piggott, J R, Sharp, R & Duncan, R E B)
London: Longman. In the press.
10. Korhola, M, Harju, K & Lehtonen, M (1989)
Fermentation. In : The Science & Technology of Whiskies.
(Eds. Piggott, J R, Sharp, R & Duncan, R E B)
London: Longman. In the press.
11. Berry, D R & Ramsay, C M (1983)
The whisky fermentation, past, present and future. pp 45 - 58.
In: Current Developments in Brewing and Distilling.
(Ed. Priest, F G & Campbell, I) London: Institute of Brewing.
12. Anderson, R G & Kirsop, B H (1974)
The control of volatile ester synthesis during the fermentation of wort of high specific gravity. J Inst Brew **80,** 48 - 55.
13. Clapperton, J F (1978)
Fatty acids contributing to caprylic flavour in beer.
The use of profile and threshold data in flavour research.
J Inst Brew. **84,** 107 - 112.
14. Watson, D C (1988), Personal communication.
15. Tressl, R, Bahri, D & Kossa, M (1980)
Formation of off-flavor components in beer.
In: The Analysis and Control of Less Desirable Flavors in Foods and Beverages. (Ed. Charalambous, G).
New York: Academic Press.
16. MacDonald, J, Reeve, P T V, Ruddleston, J D & White, F H (1984) Current approaches to brewery fermentations.
Prog Ind Microbiol **19,** 47 - 198.
17. Piggott, J R & Findlay, A J F (1984)
Detection thresholds of ester mixtures. pp. 189 - 197.
In : Flavour Research of Alcoholic Beverages.
(Eds. Nykanen, L & Lehtonen, P) Helsinki: Foundation for Biotechnical and Industrial Fermentation Research.
18. Nishimura, K & Masuda, M (1984)
Identification of some flavour characteristic compounds in alcoholic beverages.
In : Flavour Research of Alcoholic Beverages. pp. 111 - 120
(Eds. Nykanen, L & Lehtonen, P) Helsinki: Foundation for Biotechnical and Industrial Fermentation Research.

19. Leppanen, O, Ronkainen, P, Denslow, J, Laakso, R, Lindeman, A & Nykanen, I (1983) Polysulphides and thiophenes in whisky. In : Flavour of Distilled Beverages - Origin and Development. (Ed. Piggott, J R) Chichester: Ellis Horwood. pp 206 - 214.
20. Viro, M (1984)
N-heterocyclic aroma compounds in whisky. pp. 227 - 234. In : Flavour Research of Alcoholic Beverages. (Eds. Nykanen, L & Lehtonen, P) Helsinki: Foundation for Biotechnical and Industrial Fermentation Research.
21. Guymon, J F & Crowell, E A (1972)
GC separated brandy components derived from French and American oaks. Amer J Enol Viticult **23**, 114 - 120.
22. Sharp, R (1983)
Analytical techniques used in the study of whisky maturation. pp. 143 - 157 In: Current Developments in Brewing & Distilling. (Ed. Priest, F G & Campbell, I) London: Institute of Brewing.
23. Nishimura, K & Matsuyama, R (1989)
Maturation and maturation chemistry. In: The Science & Technology of Whiskies. (Eds. Piggott, J R, Sharp, R & Duncan, R E B) London: Longman.
24. Nishimura, K, Ohnishi, M, Masuda, M, Koga, K & Matsuyama, R. (1983) Reactions of wood components during maturation. In : Flavour of Distilled Beverages - Origin and Development. pp. 241 - 255 (Ed. Piggott, J R) Chichester: Ellis Horwood.
25. Nykanen, L (1984)
Aroma compounds liberated from oak chips and wooden casks by alcohol. In: Flavour Research of Alcoholic Beverages. pp. 141 - 148 (Eds. Nykanen, L & Lehtonen, P) Helsinki: Foundation for Biotechnical and Industrial Fermentation Research.
26. Reazin, G H (1984)
Chemical analysis of whisky maturation. pp. 225 - 240 In : Flavour of Distilled Beverages - Origin and Development. (Ed. Piggott, J R) Chichester: Ellis Horwood.
27. Philp, J M (1989)
Cask quality and warehouse conditions. In: The Science & Technology of Whiskies. (Eds. Piggott, J R, Sharp, R & Duncan, R E B) London: Longman.
28. Reazin, G H (1981)
Chemical mechanisms of whisky maturation. Am J Enol Vitic **32**, 283 - 289.
29. Duerr, P (1984)
Sensory analysis as a research tool. pp. 313 - 322. In : Flavour Research of Alcoholic Beverages. (Eds. Nykanen, L & Lehtonen, P) Helsinki: Foundation for Biotechnical and Industrial Fermentation Research.

14

Formation and extraction of *cis*- and *trans*-β-methyl-γ-octalactone from *Quercus alba*

Joseph A. Maga
Department of Food Science and Human Nutrition, Colorado State University, Fort Collins, Colorado 80523, USA

ABSTRACT

Heartwood and sapwood from freshly harvested American white oak were cut into 1.75 x 8 x 40 cm staves and aged for up to six years. After 0, 2, 4 and 6 years of storage, samples were solvent extracted and cis- and trans-β-methyl-γ-octalactone were separated and quantitated gas chromatographically. Six-year aged wood was charred and the lactone concentrations compared to uncharred wood. Model alcohol/wood systems were also evaluated. The influence of alcohol concentration (0, 10, 20, 40, 60%), pH (2.5, 3.5, 4.5), aging time in alcohol (0, 1, 2, 4, 8, 16, 32 months), and charring on resulting lactone concentrations were measured. Heartwood had greater amounts of oak lactones than sapwood and were found to increase 5-fold with aging. The trans form was present at approximately 10 times the amount of the cis form. Charring roughly tripled total oak lactone concentration. Oak lactone extraction was nearly linear with time and was most complete at a pH of 3.5 in 40% ethanol.

INTRODUCTION

The flavor of alcoholic beverages is due to a composite of compounds that are derived from various sources, including volatiles present in the raw ingredients; those formed during heating and fermentation; and others resulting from subsequent aging. In the case of wood-aged wines and distilled spirits, wood can provide the opportunity for both direct compound addition and precursor formation thereby making a significant contribution to product flavor complexity. Relative to oak, the wood of choice for the aging of most alcoholic beverages, the cis and trans forms of the compound β-methyl-γ-octalactone, whose structures are shown in Figure 1, are of special interest since their presence in oak and alcoholic beverages aged in oak has been documented [1-14].

trans-form *cis*-form

Figure 1. Structures of cis and trans β-methyl-γ-octalactone.

It is of interest to note that the trans form has a reported odor threshold of 0.05-0.06 ppm [9,15] while that of the cis form is somewhat higher (0.79 ppm). Also, the extent of wood aging and the overall flavor acceptability of distilled spirits has been correlated [9,10] although differing amounts of "oak lactones" or "whiskey lactones" have been found in American and European oaks [8-10]. Therefore, the objectives of this study were to more closely follow the levels of both cis and trans "oak lactones" in American white oak (*Quercus alba*) as influenced by both the length of wood aging after harvest and wood location within the tree (heartwood versus sapwood) in conjunction with the practice of charring, and to monitor the extraction of these lactones from a model system as influenced by alcohol concentration, pH and time.

MATERIALS AND METHODS

Wood source, storage and charring

Five American oak trees identified to be *Quercus alba* at least 35 years old were harvested and representative samples of heartwood and sapwood were commercially cut into 1.75 x 8 x 40 cm staves that were randomly stacked relative to individual tree source and aged in an unheated building whose temperature ranged from approximately 0-30°C depending upon time of year. The wood was aged for up to six years and three sets of heartwood and sapwood samples were randomly taken for analysis after 0, 2, 4 and 6 years. For the model system studies, samples of sapwood that had been aged for six years were manually charred on one side over an open flame to simulate commercial charring.

Model system studies

Samples of uncharred and charred six-year stored sapwood were suspended halfway into 900 ml of 0, 10, 20, 40 or 60% ethanol/water (v/v) solutions adjusted to pH 3.5 in sealed 1 L glass containers to simulate wine and distilled spirit contact with wood during aging. The vessels were stored at 18-20°C in the dark for up to 32 months and liquid samples removed after 1, 2, 4, 8, 16 and 32 months for "oak lactone" analysis. The role of solution pH on "oak lactone" extraction from both noncharred and charred sapwood stored for six years was evaluated by suspending wood samples as described above in 40% ethanol (v/v) that had been adjusted to pH 2.5, 3.5 or 4.5 using citric acid.

Oak lactone extraction and analysis

Wood samples (200g) were ground to produce sawdust which was solvent extracted as described by Onishi et al. [10] and gas chromatographically separated, identified and quantitated as described by Otsuka et al. [9]. The same procedures were used with 500 ml aliquots of solutions evaluated in the model system portion of the study.

RESULTS AND DISCUSSION

Relative to the influence of wood storage time and wood source, several obvious observations can be drawn from the data summarized in Figure 2. First, it is quite apparent that in American Quercus alba trans-β-methyl-γ-octalactone is more prevalent in both heartwood and sapwood than cis-β-methyl-γ-octalactone. Freshly harvested heartwood contained 12 mg/100 g of dry weight wood of the trans lactone while sapwood had 7/mg/100g. In contrast, the corresponding cis levels were approximately 1 mg/100 g, respectively. As a

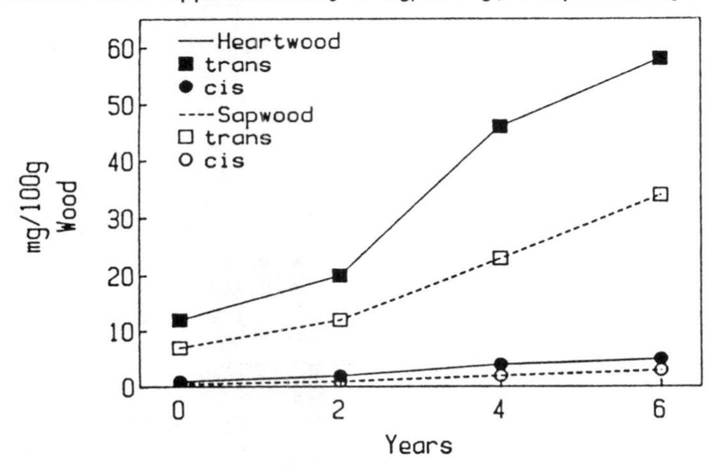

Figure 2. Influence of wood storage time on the cis and trans-β-methyl-γ-octalactone concentration (mg/100 g dry weight of wood) in the heartwood and sapwood of American Quercus alba.

result, overall, heartwood contained higher levels of oak lactones than did sapwood. Compositional differences in lipids, which can serve as precursors for lactones, between heartwood and sapwood probably accounted for this difference. The other significant observation apparent in Figure 2 is that especially with heartwood, the _trans_ lactone form increased approximately five fold going from 12 to 58 mg/100 g over the six-year aging process. Again, lipid oxidation probably accounted for this change. The _cis_ form also increased with time going from 1 to 5 mg/100 g in sapwood. Thus these data would indicate that storing wood after harvest can significantly influence the amount of oak lactones formed and thus available for extraction into alcoholic beverages.

Data presented in Table 1 clearly demonstrate the influence of charring on the formation of _cis_ and _trans_-β-methyl-γ-octalactones. In the case of heartwood, charring increased the _trans_ form from 57 to 175 mg/100 g of dry weight wood while the _cis_ form increased from 5 to 14 mg. Similar trends were noted in sapwood but initial levels were lower. Thus charring can significantly increase the amounts of oak lactones formed probably due to thermal lipid oxidation reactions.

Table 1. Influence of wood source and charring on the _cis_ and _trans_-β-methyl-γ-octalactone concentrations in American _Quercus_ _alba_ stored for six years.

Wood Source/Charred		mg/100g of Dry Weight Wood cis	trans	Total
Heartwood	Yes	14	175	189
	No	5	57	62
Sapwood	Yes	10	92	102
	No	3	35	38

The role of solution pH on the ability to extract oak lactones from noncharred and charred sapwood is summarized in Figure 3. It can be seen that over the practical pH range of most alcoholic beverages, a pH of 3.5 resulted in the highest extraction. This was especially true for charred sapwood where 9mg/L of oak lactones were found at pH 3.5, while at pH 4.5, approximately half this amount was found. The same general trend was seen for noncharred sapwood but charring produced approximately three times more total lactones. Therefore, alcoholic beverage pH as well as charring can significantly influence oak lactone amounts and extraction rates.

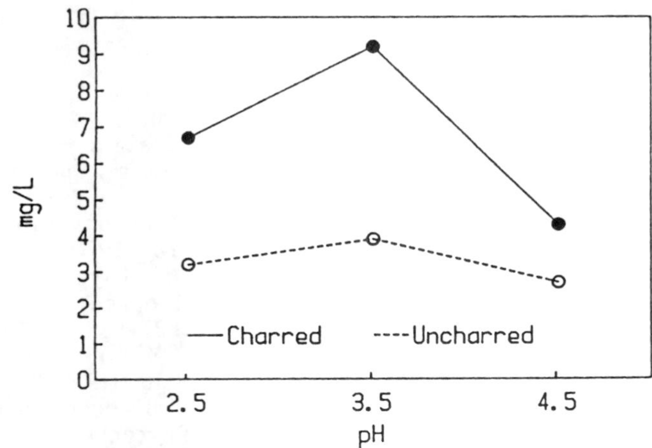

Figure 3. Influence of charring and pH on the extraction of β-methyl-γ-octalactones in wood from a 40% ethanol solution stored for 32 months.

The model system data in Figure 4 were obtained to simulate the alcohol levels in a wide range of alcoholic beverages along with typical storage times. As can be seen, small amounts of lactones were extracted in the 100% water system at pH 3.5 over the 32-month storage period. In contrast, the 40% ethanol system extracted the most lactones followed by the 60 and 20% systems. Again, charring produced higher levels of lactones than noncharred sapwood. In all systems total lactone concentration increased in a stepwise fashion with time, thereby confirming the concept that "oak lactone" concentration can serve as an indicator of the degree of wood aging for alcoholic beverages.

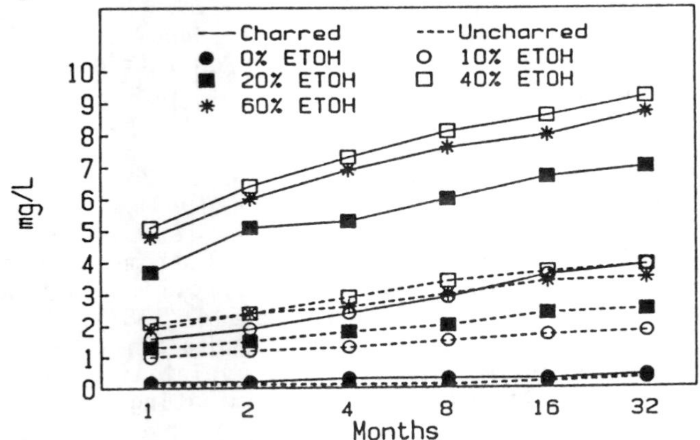

Figure 4. Influence of charring, alcohol concentration and time on the extraction of β-methyl-γ-octalactones from sapwood in a model system at pH 3.5.

REFERENCES

1. Otsuka, K., S. Imai, and M. Sambe. Agric. Biol. Chem. 30:1191 (1966).
2. Kahn, J.H., P.A. Shipley, E.G. LaRoe, and H.A. Conner. Whiskey composition: Identification of additional components by gas chromatography-mass spectrometry. J. Food Sci. 34:587-591 (1969).
3. Suomalainen, H., and L. Nykanen. Composition of whisky flavour. Process. Biochem. 5:13-18 (1970).
4. Liebich, H.M., W.A. Koenig, and E. Bayer. Analysis of the flavor of rum by gas chromatography and mass spectrometry. J. Chromatog. Sci. 8:527-533 (1970).
5. Nishimura, K., and M. Masuda. Minor constituents of whisky fusel oils. J. Food Sci. 36:819-822 (1971).
6. Masuda, M., and K. Nishimura. Branched nonalactones from some _Quercus_ species. Phytochem. 10:1401-1402 (1971).
7. Kepner, R.E., A.D. Webb, and C.J. Muller. Identification of 4-hydroxy-3-methyloctanoic acid gamma-lactone [5-butyl-4-methyldihydro-2-(3H)-furanone] as a volatile component of oak-wood-aged wines of _Vitis_ Vinifera var. "Cabernet Sauvignon". Am. J. Enol. Vitic. 23:103-105 (1972).
8. Guymon, J.F., and E.A. Crowell. GC-separated brandy components derived from French and American oaks. Am. J. Enol. Vitic. 23:114-120 (1972).
9. Otsuka, K., Y. Zenibayashi, M. Itoh, and A. Totsuka. Presence and significance of two diastereomers of β-methyl-γ-octalactone in aged distilled liquors. Agric. Biol. Chem. 38:485-490 (1974).
10. Onishi, M., J.F. Guymon, and E.A. Crowell. Changes in some volatile constituents of brandy during aging. Am. J. Enol. Vitic. 28:152-158 (1977).
11. Feuillat, M., M.A. Radix, D. Dubois, and J. Dekimpe. Elevage des vins de Bourgogne en futs de chene. Vignes Vins 299:5-10 (1981).
12. Feuillat, M. L'elevage des vins en futs de chene. Industr. Bevande 68:463-472 (1983).
13. Naudin, R. L'elevage des vins de Bourgogne en futs de chene. Vignes Vins 324:19-33 (1983).
14. Gunther, C., and A. Mosandl. Stereoisomeric aroma XV. chirospecific aroma of natural aroma components: 3-methyl-4-octanolide-"Quercus-, whiskylactone." Z. Lebensm. Unters. Forsch. 185:1-4 (1987).
15. Salo, P., L-Nykanen, and H. Suomalainen. Odor thresholds and relative intensities of volatile aroma components in an artificial beverage imitating whisky. J. Food Sci. 37:394-398 (1972).

15

The distribution of lignin breakdown products through new and used cask staves

John Conner, Alistair Paterson and John Piggott
Food Science Division, Department of Bioscience and Biotechnology,
University of Strathclyde, 131 Albion Street, Glasgow, G1 1SD, Scotland

INTRODUCTION

Maturation is the final step in the whisky production process with the fresh distillate being kept in oak casks for a minimum of 3 years in Scotland. During this period the new distillate becomes highly modified as a result of its contact with the wood. The interactions that occur between the cask wood and the fresh distillate have been described as additive, subtractive or interactive [1]. One of the additive interactions that occurs during maturation is the direct extraction of wood components from the cask. Of the extracted components the degradation products of lignin make a significant contribution to the maturation process. Four pathways for the presence of these lignin related compounds in mature spirits have been identified [2]. These involve anaerobic degradation of lignin to aromatics when the oak cask is toasted or charred, extraction of monomeric compounds and native lignin by the spirit, ethanolysis of lignin and further conversion of compounds in the spirit.

It is clear that the type of cask used can have a major effect on the maturity of the final product. In the Scotch whisky industry casks are reused many times until they become exhausted, i.e. fail to provide a satisfactory maturation. Attempts to rejuvenate used casks by scraping and recharring have been studied [3] and though this did lead to an increase in the levels of congeners depleted by previous use, the amount did not equal that found in a new cask.

During maturation ethanol and water penetrate the entire depth of the cask wood but it is not known to what depth and degree the formation and extraction of lignin breakdown products occur. An analysis of the aromatic aldehyde content of the inner and outer faces of staves that had contained Armagnac for 20 years found the inner face was 16 times richer in such compounds than the outer surface [4].

MATERIALS AND METHODS

The three staves used in this study came from a used bourbon cask, a first fill Scotch whisky cask and an exhausted cask.

For each stave a 5g sample of wood was scraped from the char layer (depth C in Figures) and then at a depth of 5, 10, 15, 20 and 25 mm through the cask stave. The wood shavings were then extracted with $CHCl_3$ using a Soxhlet apparatus, dried and redissolved in 60% ethanol and filtered through a Bond Elut C18 column.

Colour was determined from the absorbance at 430 nm in a 1 cm quartz cell and calculated by the equation

$$\text{Whisky Colour} = 10 \times A_{430} \times 1.27$$

and relates to 1 gram of wood.

Total phenolic content was determined using Folin Dennis reagent and standard solutions of gallic acid. Quantities are expressed in milligrams of gallic acid equivalents per gram of wood.

The lignin derived compounds syringaldehyde, syringic acid, acetovanillone, vanillin and vanillic acid were identified and quantified using HPLC. A Spherisorb ODS 5 µm column 25 cm long was used. The gradient elution followed that detailed by Casteele et al. [5] using A, formic acid - water (5:95) and B, methanol. The elution profile was: 0-2 min, 7% B in A (isocratic); 2-8 min, 7-15% B in A (linear gradient); 8-25 min, 15-75% B in A (linear gradient); 25-27 min, 75-80% B in A (linear gradient); 27-29 min, 80% B in A (isocratic). A UV detector set at 280 nm linked to a TRIO integrator was used for peak detection and quantification. 2,3-Dihydroxybenzaldehyde was used as an internal standard. The concentrations of the compounds were determined from the HPLC peak areas using standard curves and are given in micrograms per gram of wood.

RESULTS AND DISCUSSION

The results of these investigations are shown graphically in Figures 1 to 7. In analysing and discussing these graphs no direct quantitative comparisons between staves was made as the wood in each was of a different age and any differences in the original wood were not known. What was thought important was the occurrence of concentration peaks and their position in the stave. For the compounds studied these peaks are thought to represent areas of their formation from breakdown of lignin.

Considering the staves in order of increased cask usage, three distinct patterns of colour and congener depletion become apparent. The first and commonest of these is shown by colour (Figure 1), total phenols (Figure 2) and the aromatic aldehydes (Figures 3 and 4). All of these showed large peaks at 10 mm depth in the used bourbon cask which persisted in the first fill stave for all but syringaldehyde. In the exhausted stave the peak is shifted to the 20 mm layer, again for all but syringaldehyde. In the case of syringaldehyde the highest concentration in the first fill stave is in the 20 mm layer with only very minor concentrations being detected in the exhausted stave. It appeared that the main pool of these constituents for extraction into the maturing whisky retreats deeper into the stave with increased cask usage. This process occurred faster in the case of syringaldehyde than for the others. In the breakdown of lignin syringyl residues are split off from the polymer more readily than the frequently more highly condensed guaiacyl or hydroxyphenyl groups [6], and from these results this applies to lignin breakdown in whisky casks as well.

The second pattern of depletion applied to the aromatic acids (Figures 5 and 6). For these low concentrations were found in the used bourbon stave except for a peak in the 10 mm layer for vanillic acid. Both then occurred in high concentrations in the outside layer of the first fill stave with very much smaller concentrations being found in the inner layers. In the exhausted cask this appeared to be reversed most noticeably for vanillic acid where the highest concentration was found in the inside layer with none at all detected in the outer 3 layers. These compounds appear to be exhausted in the opposite direction to the above, from the outside of the barrel inward. This may be explained by their highly oxidised nature, their formation from lignin requiring aerial oxidation. The precursors involved in this reaction appear to be depleted in the course of cask exhaustion from the outside face of the cask in contact with air to the inside face.

The third observed pattern was for acetovanillone (Figure 7). In the used bourbon stave the highest concentration was in the outer layers, for the first fill stave this had shifted to the inner layers and for the exhausted stave it had shifted back to the outer layers. Thus for the first fill and exhausted cask the pattern is essentially the reverse of the second one described previously. Formation and extraction of acetovanillone appear to proceed from the inside of the barrel to the outside. This compound is the product of the ethanolysis of oak-wood lignin in the whisky cask. From this study it appears that the lignin involved in this reaction is first broken down on the inner faces of the cask working its way to the outside and cask exhaustion.

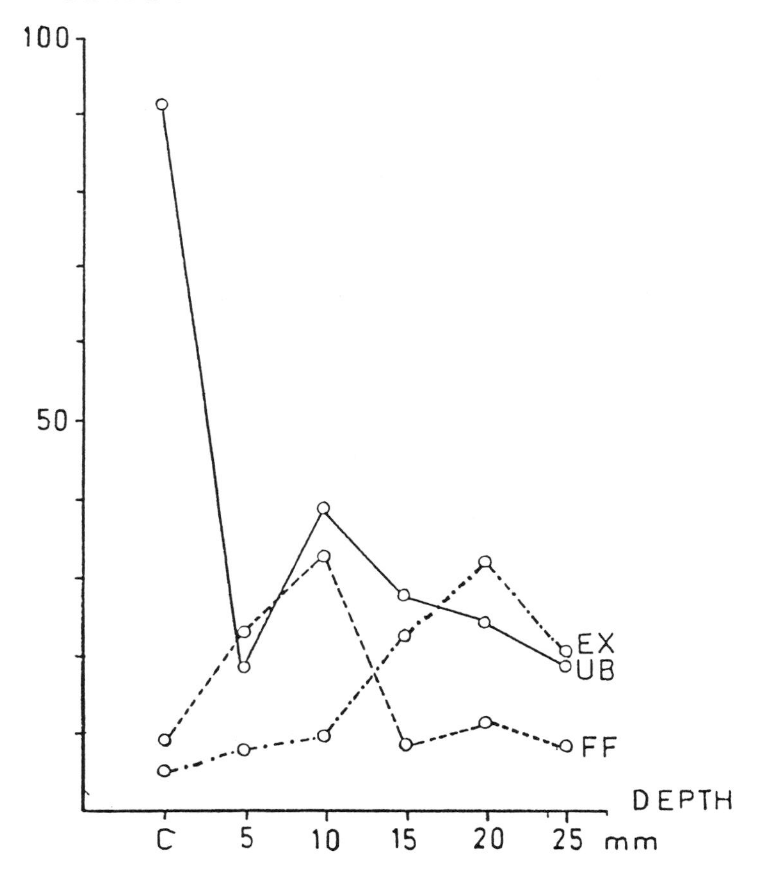

Figure 1: Plot of extracted colour units per 5g of wood against wood sample depth for used bourbon (UB), first fill (FF) and exhausted (EX) cask staves.

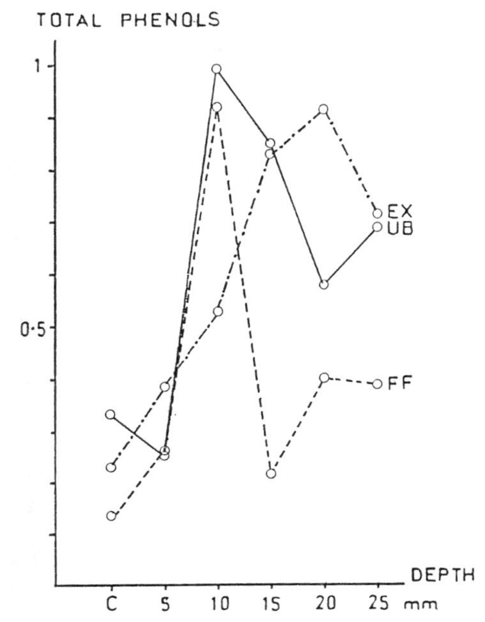

Figure 2: Plot of total phenols (expressed in mg of gallic acid equivalents per gram of wood) against wood sample depth (mm) for used bourbon (UB), first fill (FF) and exhausted (EX) cask staves.

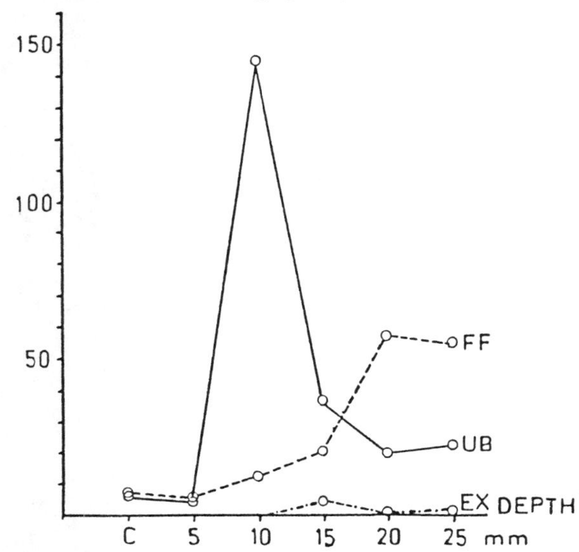

Figure 3: Plot of syringaldehyde concentrations (μg per gram wood) against wood sample depth (mm) for used bourbon (UB), first fill (FF) and exhausted (EX) cask staves.

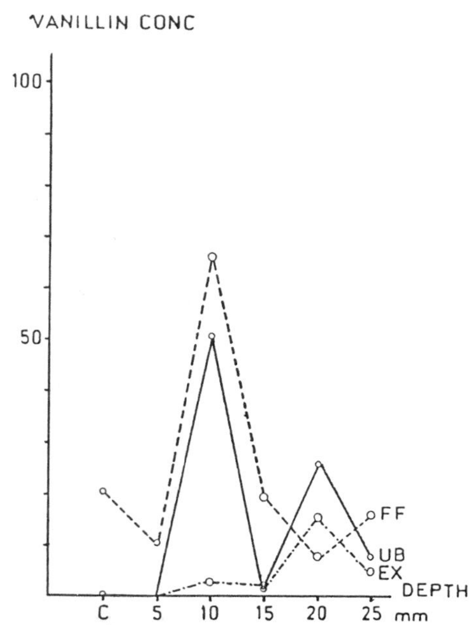

Figure 4: Plot of vanillin concentrations (μg per gram wood) against wood sample depth (mm) for used bourbon (UB), first fill (FF) and exhausted (EX) cask staves.

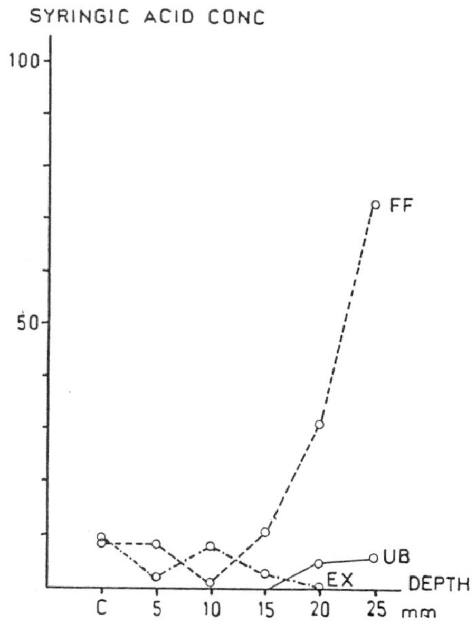

Figure 5: Plot of syringic acid concentrations (μg per gram wood) against wood sample depth (mm) for used bourbon (UB), first fill (FF) and exhausted (EX) cask staves.

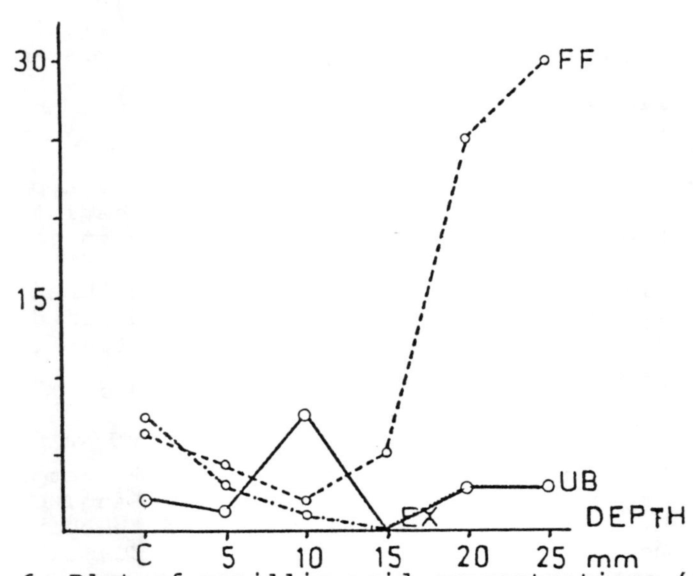

Figure 6: Plot of vanillic acid concentrations (μg per gram wood) against wood sample depth (mm) for used bourbon (UB), first fill (FF) and exhausted (EX) cask staves.

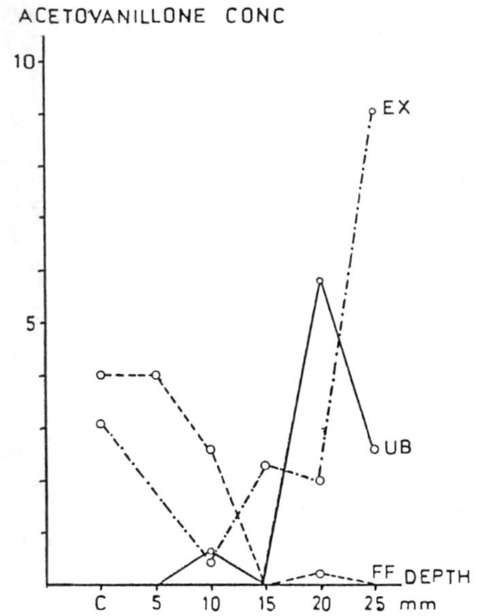

Figure 7: Plot of acetovanillone concentrations (μg per gram wood) against wood sample depth (mm) for used bourbon (UB), first fill (FF) and exhausted (EX) cask staves.

CONCLUSIONS

With increased cask usage the highest levels of colour, total phenols, syringaldehyde and vanillin occurred deeper in the cask stave. Syringaldehyde was observed to be depleted faster than vanillin. Two other types of reaction involving lignin were noted:

i) Ethanolysis, represented by the appearance of acetovanillone, progressed from the inner surface of the first fill stave to the outer surface of the exhausted stave.

ii) Oxidation, represented by the appearance of syringic and vanillic acids, progressed from the outer surface of the first fill stave to the inner surface of the exhausted stave.

ACKNOWLEDGEMENTS

The authors are grateful to Dr. D. Watson, Chivas Brothers Ltd., Keith and Mr. J. Lang, Chivas Brothers Ltd., Paisley for the supply of the barrel staves used in this study. One of us (JC) acknowledges the award of a research fellowship from the A.F.R.C..

REFERENCES

1. Philp, J., 1988. Cask Quality and Warehouse Conditions. In Piggott, J.R., Sharp, R., and Duncan, R.E.B. (eds.) The Science and Technology of Whiskies. Longman, London.
2. Nishimura, K., Ohnishi, M., Masuda, M., Koga, K. and Matsuyama, R., 1983. Reactions of wood components during maturation. In Piggott, J.R. (ed.) Flavour of Distilled Beverages: Origin and Development. Ellis Horwood, Chichester.
3. Reazin, G.H., 1981. Chemical mechanisms of whisky maturation. Am. J. Enol Vitic. **32**, 283-89.
4. Peuch, J-L., 1981. Extraction and evolution of lignin products in Armagnac matured in oak. Am. J. Enol. Vitic. **32**, 111-14.
5. Casteele, K.V., Geiger, H. and van Sumere, C.F., 1983. Separation of phenolics and coumarins by reversed-phase high performance liquid chromatography. J. Chromatog. **258**, 111-124.
6. Freudenberg, K. and Sidhu, G.S., 1961. Zur kenntnis des lignins der Buche und der Fichte. Holzforschung **15**, 33-39.

16

GC profiles and malt whisky flavour

J. E. Carter-Tijmstra
United Distillers plc, Glenochil Research Station, Menstrie, Clackmannanshire, Scotland

INTRODUCTION

Many distinctive flavours other than the '**peaty**' odour brought about by the peating of malt can be identified by nose in Scotch malt whisky. The correlation of analytical data with these flavours is desirable because it can lead to better understanding and control of spirit quality. Such correlations might be approached in two ways.

1. Measurement of the concentrations of specific congeners which are known to produce specific whisky flavours, by gas or liquid chromatography, or mass spectrometry.

2. The examination of analytical data from whiskies with known flavours to find relationships between flavour and the analytical data. Data such as the peaks from gas chromatographic (GC) traces may be useful in this search.

When the congener or congeners responsible for a flavour are unknown, the second approach may yield useful information which can lead to the identification of flavour constituents.

METHODS

Extraction

Whisky (20 ml), 20 ml water, 1 ml hexane (containing dibromopropane as internal standard), plus 0.1 ml methyl octanoate solution (internal standard) were shaken in a stoppered test-tube for 10 minutes on a mechanical shaker. Water (35 ml) was added and the tube inverted twice. After settling, the hexane layer was removed with a Pasteur pipette and stored in a sealed vial.

Gas chromatography

Column:	25 metre non-polar (SGE 25QC3BP1-1.0)
Carrier gas:	Helium
Make-up gas:	Nitrogen
Injection:	1 µl Grob splitless, extract coinjected with 1 µl alkane solution.
Oven:	5 min at 60 °C, rising at 5 °C/min to 227 °C.

RESULTS AND DISCUSSION

GC traces of extracts of mature malt whiskies that had been assessed by nose to have **sour, fruity, grassy, heavy** or **sulphury** have been examined for peaks that could be related to these flavours. Flame ionisation (FID), electron capture (ECD) and flame photometric (FPD) GC detectors were used. GC traces from extracts of these whiskies are shown together with charts relating the occurrence and size of peaks to flavour. Peak sizes are given in arbitrary units (peak area/internal standard peak area). Peaks have been identified by the Retention Index (RI), calculated from the retention times of the added alkanes on the FID trace. These alkane peaks were also used to calculate RI's on the ECD trace, there being virtually no difference in retention times between the two detectors when mounted in tandem. It is not suggested that the GC peaks shown in this work are necessarily responsible for the flavours, but they might be useful analytical markers for particular flavours or help in the search for their origin.

In malt whisky, ethyl butyrate concentrations are correlated with those of n-butyric acid [1]. This acid gives rise to a **sour** odour and is due to bacterial contamination of the fermentation. The amount of ester may therefore be used as a measure of this odour. Figure 1 shows an FID trace from a **sour** whisky and one from a normal whisky. In Figure 2 the distribution of ethyl butyrate peak sizes is shown for **sour** and normal whisky. A particular type of **fruity** odour is found in some malt whiskies and may be cask-related. FID traces from **fruity** whisky consistently showed a peak with RI 1920 (Figure 3), which was usually absent in other whiskies. Figure 4 shows the distribution of this peak in **fruity** and **non-fruity** whiskies. The ECD trace from a **grassy** malt whisky contained a peak at RI 1955 that was usually smaller or absent in other whiskies (Figures 5 and 6). An ECD peak at RI 1036 was more likely to be found in **heavy** than in other whiskies (Figures 7 and 8). In whiskies with a strong **sulphury** odour, a very large and unquantifiable peak was found in the ECD trace (Figure 9). Mass spectrometry identified this peak as sulphur. An FPD trace from **sulphury** whisky shows a large peak at the same retention time to that on the ECD trace (Figure 10).

The results presented here demonstrate some of the possibilities of this approach, particularly that of using GC detectors such as the ECD which have a different range of sensitivities to that of the more commonly used FID.

REFERENCE

1. Carter-Tijmstra, J.E., Proceedings of the Second Aviemore Conference on Malting, Brewing & Distilling. Institute of Brewing, London, 413, 1986.

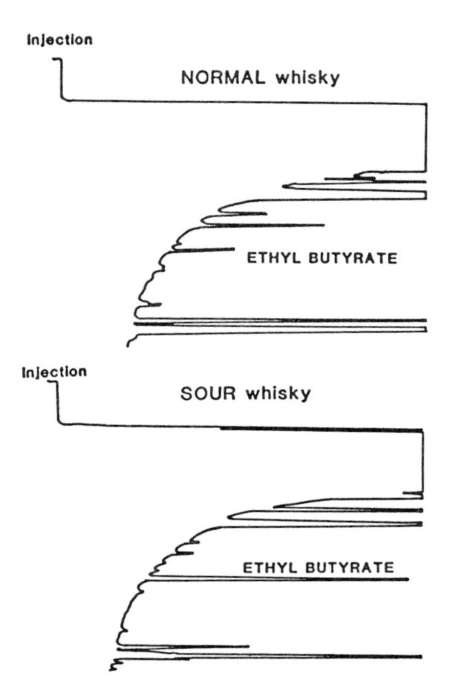

Figure 1. Part of FID Trace Showing Ethyl Butyrate Peak

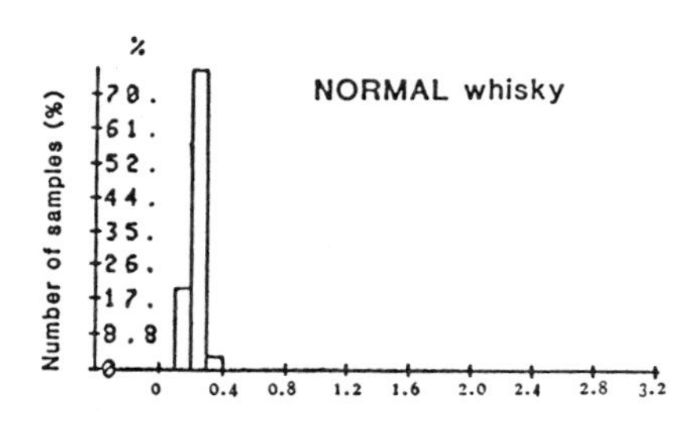

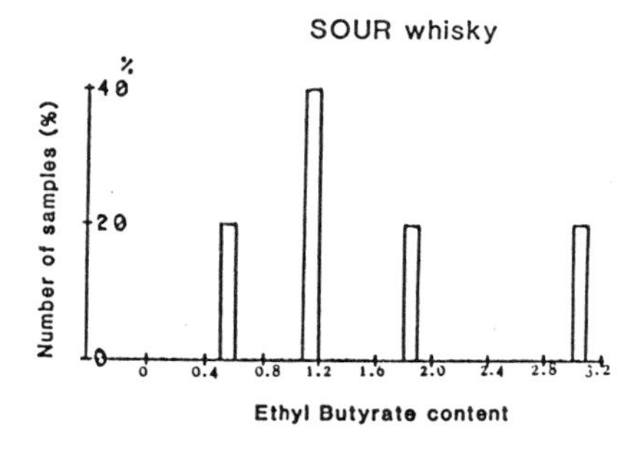

Figure 2. Ethyl Butyrate in Normal and SOUR Whisky

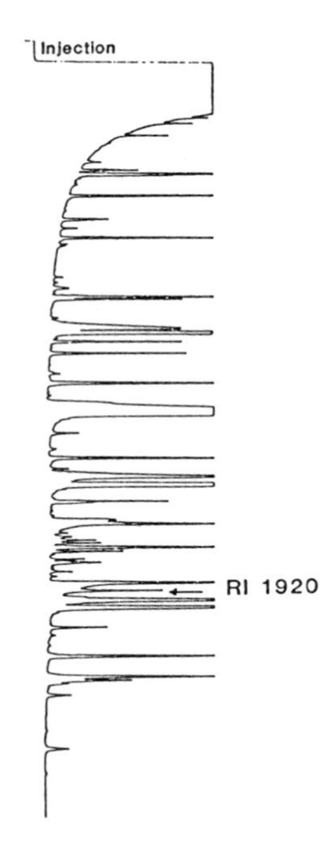

Figure 3. FID Trace from FRUITY Whisky

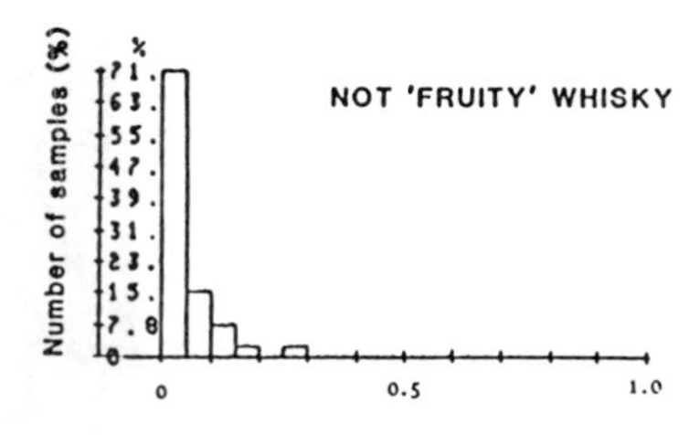

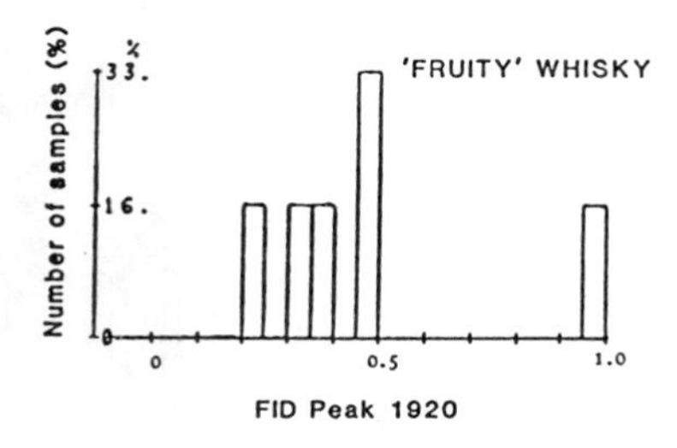

Figure 4. FID Peak 1920 in FRUITY Whisky

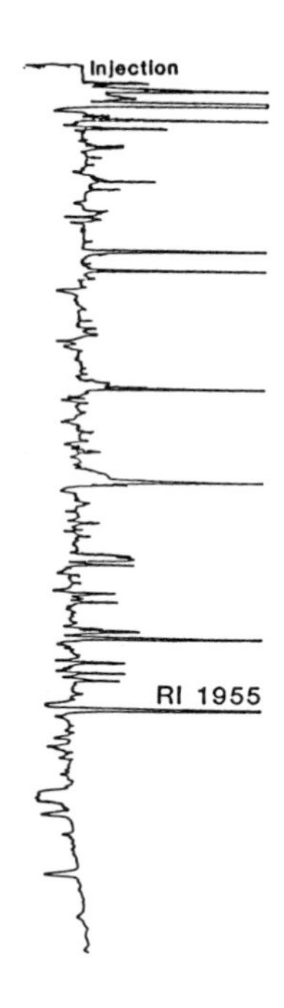

Figure 5. ECD Trace from GRASSY Whisky

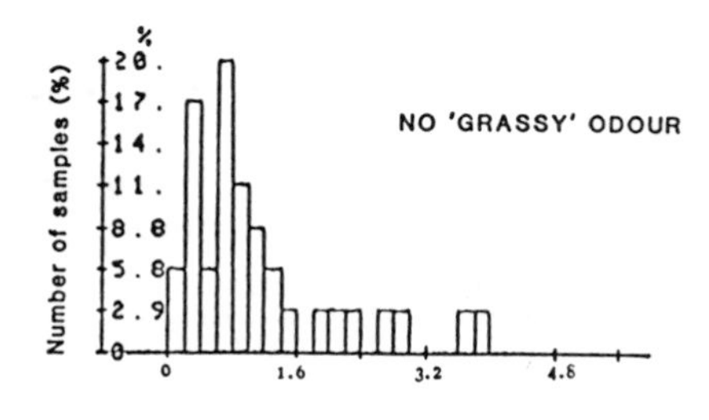

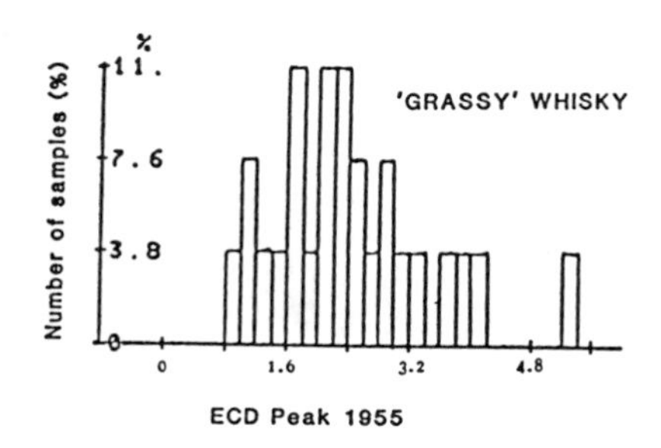

Figure 6. ECD Peak 1955 in GRASSY whisky

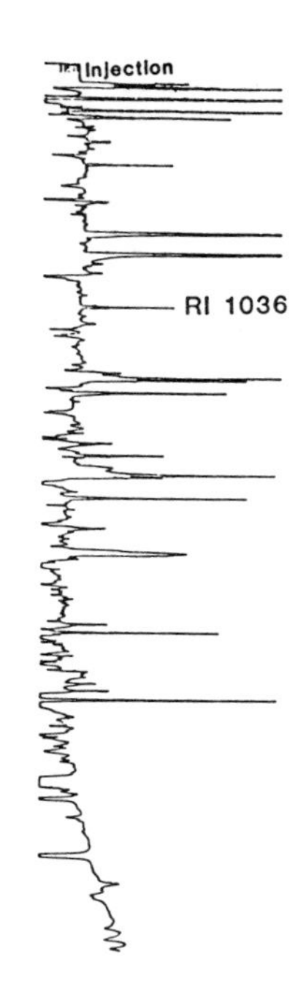

Figure 7. ECD Trace from HEAVY Whisky

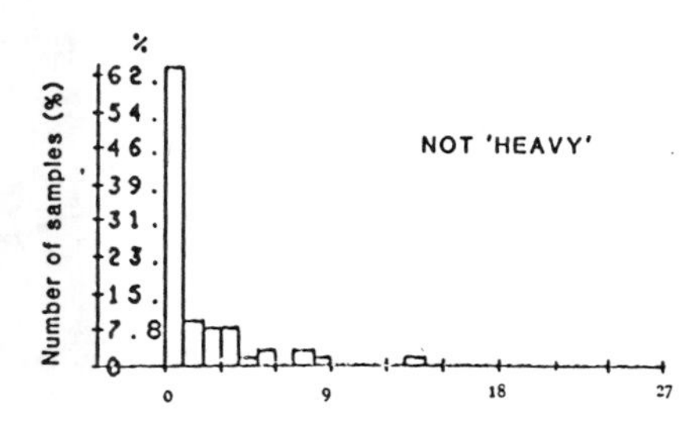

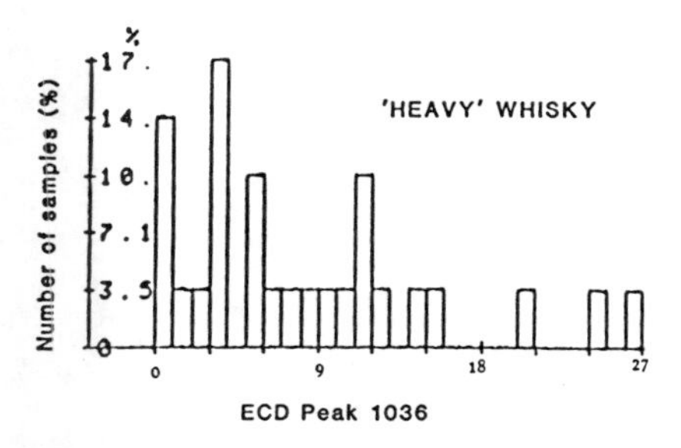

Figure 8. ECD Peak 1036 in HEAVY Whisky

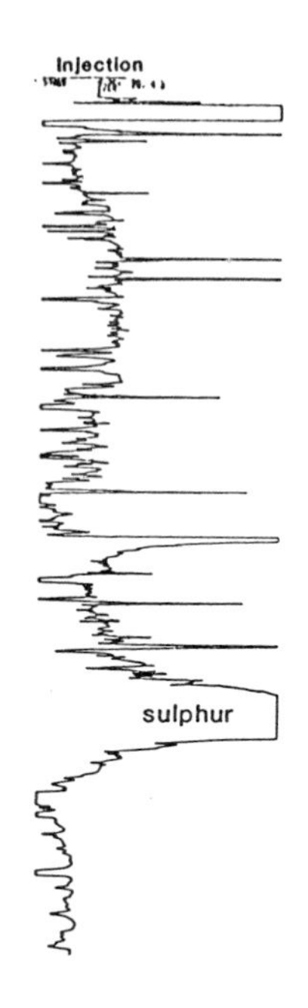

Figure 9. ECD Trace from SULPHURY Whisky

Figure 10. FPD Trace from SULPHURY Whisky

17

Influence of lactic acid bacteria on aldehyde, ester and higher alcohol formation during Scotch whisky fermentations

Pamela A. Geddes and Harry L. Riffkin
Pentlands Scotch Whisky Research Ltd., 84 Slateford Road, Edinburgh, EH11 1QU, Scotland

INTRODUCTION

Scotch Whisky fermentations in both malt and grain distilleries are brought about by non-sterile, self-regulating mixed microbial cultures. A number of parameters limit microbial growth. These parameters include pH, alcohol concentration and availability of carbohydrate and nitrogenous material. The added yeast, *Saccharomyces cerevisiae*, dominates the flora, but other yeasts and bacteria enter the system via raw materials (malt, yeast and process water) and the plant itself (Dolan, 1979). The survival of most genera such as *Enterobacter*, *Bacillus* and *Micrococcus* is short-lived due to their low tolerance of acid and alcohol. However, lactic acid bacteria are well suited to the distillery environment.

The secondary fermentation by species of *Lactobacillus*, following completion of yeast fermentation, has been shown to be an integral feature of distillery fermentations (Geddes, 1987). This secondary fermentation does not adversely affect spirit yield and is believed to make a positive contribution to spirit quality and flavour, although the parameters of this contribution are not yet understood.

During the secondary fermentation the lactic acid bacteria utilise substrates which are non-fermentable by yeast including residual dextrins and pentose sugars. Various metabolites are produced, of which the most abundant are acetic and lactic acids. These compounds may be chemically converted during the subsequent distillation process.

This work demonstrates the influence of lactobacilli on aldehyde, ester and higher alcohol formation during Scotch Whisky production. Sensory evaluation of new-make spirits, derived from laboratory fermentations where varying levels of lactobacillus were present, has been undertaken.

METHOD

A method of producing bacteria-free distillers all-malt wort has been developed. Unboiled laboratory malt wort, with an original gravity of about 1055, is filtered while still warm using prefilter pads in combination (Millipore - sizes AP25 and AP15). The wort is then re-heated to 64.5°C and held at this temperature for 1 h. Following re-heating, the wort is cooled to 25°C and aliquots are drawn off as required into sterile fermentation vessels.

Laboratory cultured single strain (bacteria-free) distillery yeast is then added in suitable quantities (0.4 per cent w/v) to the worts. The combination of these techniques ensures that the fermenting wort is essentially free of microbial contamination from the usual sources (i.e. malt and yeast).

Specific cultures of lactobacilli are introduced after 24 h or 48 h fermentation. Fermentation times are prolonged and fermentation temperatures can be varied. Subsequent distillations for low wines and spirit are carried out in laboratory copper pot stills.

RESULTS AND DISCUSSION

In the first experiment, new-make distillates were obtained from 7 defined fermentations after 168 h. All samples (2 l) were of bacteria-free wort pitched with laboratory-produced DCL 'M' type yeast. Samples 1,5 and 7 were inoculated with lactobacilli after 24 h fermentation while samples 2,4 and 6 were left to ferment without the addition of lactic acid bacteria. Sample 3 was similar to sample 2 except that nisin was added to prevent the growth of any lactic acid bacteria which may have survived the combined sterilisation procedures.

Samples 1,2 and 3 were fermented at 30°C, 4 and 5 at 33°C and 6 and 7 at 27°C. Distillation rates were 3.5 ml per min for the wash distillation and 2.5 ml per min for the spirit distillation.

Fig 1 shows the pattern of lactobacillus growth in samples 1-7 and this bar chart clearly emphasises the degree of sterility in samples 2,3,4 and 6 and the high level of added infection

Figure 1 Lactobacillus populations in samples 1 - 7 prior to distillation

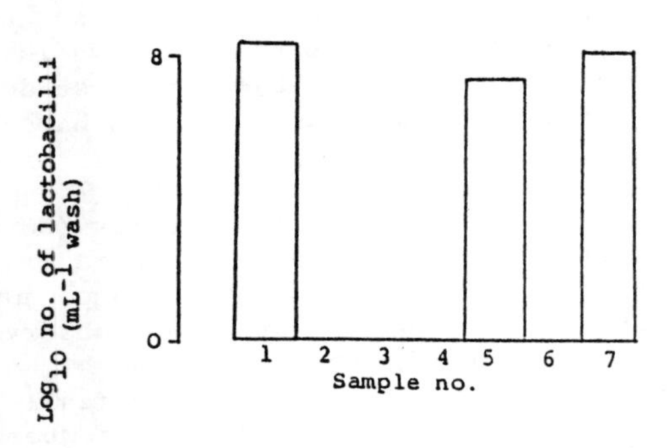

Figure 2 Levels of acetaldehyde, fatty acids and their ethyl esters in spirits from samples 1 - 7

(a) acetaldehyde

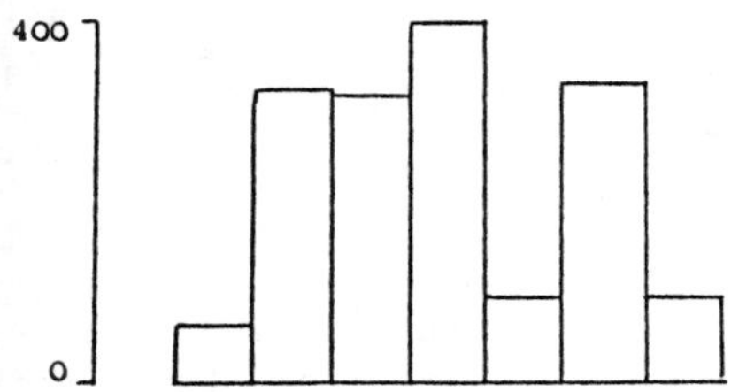

(b) fatty acids and their ethyl esters

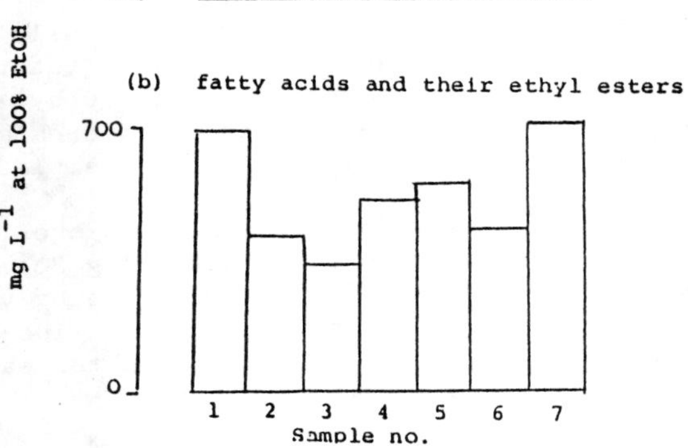

in samples 1,5 and 7.

Acetaldehyde was analysed because it has a low sensory threshold and a significant role in affecting whisky flavour. Very large differences between the inoculated and un-inoculated products were apparent and these are shown in fig 2a. During maturation, levels of acetaldehyde increase due to oxidation of ethanol, consequently it is desirable to have low levels in new-make spirit.

The results for the levels of fatty acids (C6 and over) and their associated ethyl esters are shown in fig 2b. Sample 1 contained significantly more of these compounds than did samples 2 and 3. The same pattern was observed with respect to samples 7 and 6 and may be due to the higher acidity of inoculated wash which makes it more chemically reactive on distillation. The higher fermentation temperature (33°C) may have stimulated the biosynthesis of fatty acids by yeast and suppressed their production by lactobacilli.

The small chain esters of ethyl lactate and ethyl acetate were also determined by GC and the data are shown in figs 3a and 3b. The levels of ethyl lactate differed significantly within sample sets. Not surprisingly the level of this compound was much greater in sample 1 which had undergone a strong secondary acid-producing fermentation. Samples 5 and 7 prior to distillation yielded high numbers of lactobacilli but these bacteria did not produce much acid.

Sample 1 contained the highest amount of acetic acid since 1 of the 2 strains of lactobacilli inoculated was strongly heterofermentative, yet its level of ethyl acetate is the lowest. However, since esters in wash are not usually formed by direct esterification - it could be that, since yeast autolysis occurred fairly early in sample 1, less ethyl acetate was synthesised by yeast. No significant differences were apparent between the levels of ethyl acetate in samples 4 - 7.

There were essentially no differences in the levels of higher alcohols recorded between the 3 sets (fig 3c). Since higher alcohols are formed within the first 24 h by yeast from amino acids, these results were not surprising since every fermentation was identical to its duplicate in the early stages prior to lactobacillus inoculation at 24 h.

The previous experiments were carried out using exclusively DCL 'M' type. However this does not reproduce exactly the conditions in many malt distilleries where wort is fermented by mixtures of distillers and brewers yeast. Additionally, distillery fermentations are not attemperated. Accordingly, a further series of experiments was undertaken to investigate the effect of mixed yeast cultures, fermenting under simulated

Figure 3 Levels of ethyl lactate, ethyl acetate and higher alcohols in spirits from samples 1 - 7

(a) ethyl lactate

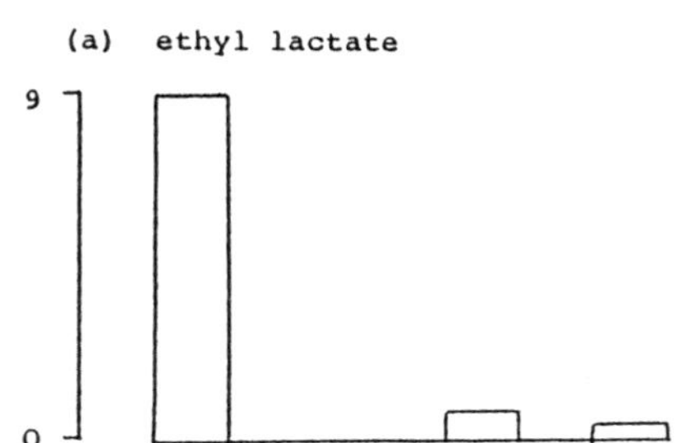

(b) ethyl acetate

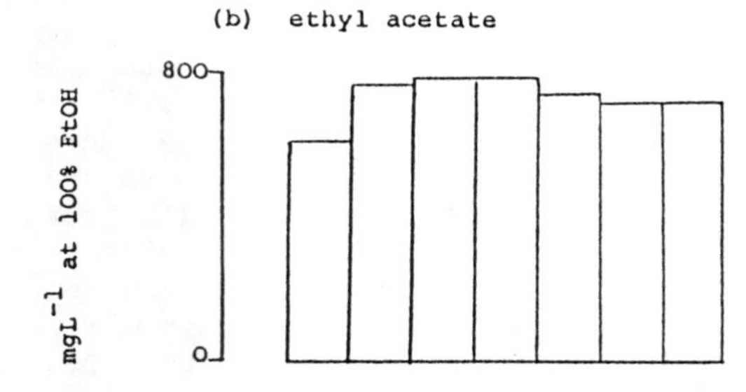

(c) higher alcohols

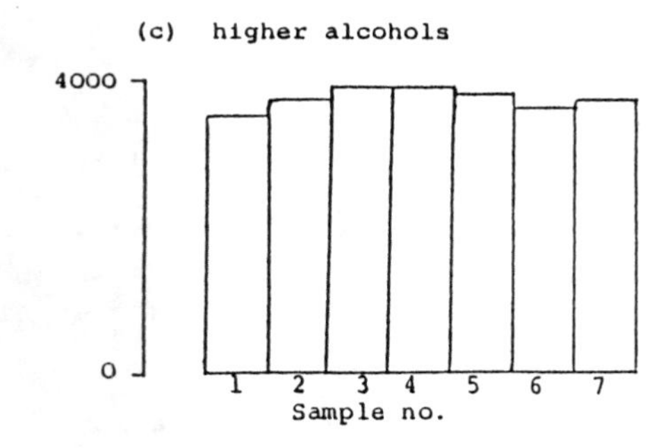

distillery conditions, on the production of acids and esters.

The experimental design was essentially similar to that previously described. Four samples of fermented wash (4 l) were obtained by the addition of (a) DCL 'M' type yeast (samples 1 and 2) and (b) DCL 'M' type and brewers yeast (samples 3 and 4). Both types of yeast were pure cultures and fermentations were carried out in Thermos cool jugs (8 l capacity) which simulate distillery washback conditions. One of each duplicate set was inoculated with a mixed culture of lactobacilli after 24 h (samples 2 and 4). The levels of fatty acids and esters were measured in the resultant spirits by GC (data not shown) and sensory evaluations were carried out on the spirits by a trained panel.

Both inoculated samples yielded significantly higher levels of ethyl caprate, ethyl laurate and ethyl lactate than their uninoculated counterparts. The total fatty acid esters were also higher in the former.

The sensory panel preferred the spirits from samples 3 and 4 to those from samples 1 and 2. Additionally, the spirit from sample 2, which had undergone late lactobacillus fermentation, was preferred to the spirit from sample 1. The basis for preference was a higher level of estery character. It was interesting to note that the latter spirit was described by one of the expert panellists as having a synthetic aroma and a non-whisky note. The verdicts on spirits from samples 3 and 4 suggested a marginal preference for the spirit from sample 4. However, both of these distillates were more complex and this complexity is believed to be an essential factor in the subsequent maturation process.

CONCLUSIONS

The presence of lactobacilli in controlled fermentations led to reduced levels of acetaldehyde and higher levels of fatty acids and their ethyl esters in the resulting distillates. The levels of ethyl lactate in the distillates were also elevated when lactobacilli were present. The secondary lactobacillus fermentation had no apparent influence on higher alcohol formation.

The effect of the secondary fermentation on the quality of the distillates was assessed by sensory analysis. The data led to the conclusion that the presence of lactobacilli was essential for the development of the perceived quality parameters of Scotch malt spirit.

REFERENCES

Dolan, T.C.S. (1979). Bacteria in whisky production.

The Brewers 65, 60-64.

Geddes, P.A. (1987). Growth of lactobacilli in Scotch Whisky fermentations. M.Sc. Thesis, Heriot-Watt University, Edinburgh.

18

Odour intensities of whisky compounds

D. R. Perry
United Distillers plc, Glenochil Research Station, Menstrie, Clackmannanshire, Scotland

ABSTRACT

Whisky contains a large number of aroma compounds, only some of which make a noticeable contribution to the bouquet. Odour intensities above the threshold concentration increase at different rates, according to the value of an intensity index. This index has been determined for eight whisky congeners, and the significance of their odour contributions assessed. Esters have the highest and carbonyl compounds the lowest indices. Acetal is a major contributor to mature whisky odour.

INTRODUCTION

The threshold concentration (t) of an aroma compound is the lowest at which the aroma may be detected. The intensity of the aroma increases at higher concentrations at a rate specific to that compound. An odour unit (o.u.) may be defined as the ratio of the concentration (c) to the threshold concentration (i.e. threshold multiple c/t).

Aroma compounds are present in whisky in amounts up to several thousand odour units. But some odours are more intense than others even when the compounds responsible are present at the same odour unit concentration. Odour threshold concentrations and intensities at higher concentrations have been measured for eight whisky compounds so that their contribution to whisky aroma may be assessed. The range of different odours present can be appreciated by nosing stepwise dilutions of whisky.

THE AROMA OF DILUTED SPIRITS

Two malt spirits were diluted as follows. Water was added to 10 ml of spirit so that the final ethanol content was 23%v., and 10 ml of this solution was diluted further to 30 ml with 23%v. ethanol. This last process was repeated successively to give a range of dilutions:

3 : 9 : 27 : 81 : 243 : 729 : 2187

The amount of congener was reduced in each dilution but the ethanol content remained constant. The solutions were presented in nosing glasses to experienced panellists. They were asked to nose each solution and describe the aroma. The spirits chosen were:

(a) Newly made malt spirit sampled during cask filling.
(b) 12-year-old malt whisky sampled from a sherry butt.

The results are shown in Table 1.

Table 1. Aroma descriptions of diluted spirit

Dilution	New malt spirit	12 y.o. malt whisky
3	bran	mature/very pleasant
9	bran/malty	fruity
27	malty/feints	vanillic/feints
81	sour/feints	perfumed/yeasty
243	sour	strong fragrance
729	fragrant/sour	fragrant
2187	weak fragrance	weak fragrance

The odour intensity decreased and the character of the aroma changed in each successive dilution. This observation is not surprising. The odour contribution of a compound present at ten times its threshold concentration will no longer be noticed when diluted ten-fold. But the dilution experiment also shows that the fragrant compound is not perceived in the less dilute samples. The fragrant aroma is masked by odours of compounds present at smaller odour unit concentrations.

The odour of the most dilute spirits is described as a weak, sweet fragrance. The compound responsible for that odour was identified as ß-damascenone. This compound has been previously identified as the principal odorant of Damascus rose oil [1].

Damascenone has a low threshold concentration of 0.00004 mg/litre (in 23%v. ethanol). It is present in new spirit and maturation strength whisky (63%v. ethanol) at 0.1 mg/litre (2500 o.u.). The odour perceived in diluted spirit has a pure quality, suggesting that no other compound is present at a greater odour unit concentration. However, the odour of whisky is definitely not like the odour of damascenone: this is apparent from the odour descriptions of diluted spirits.

The odour intensity of damascenone is low compared to the intensities of the compounds responsible for the masking odours. The odour contribution of a compound can be assessed by estimating the odour intensities at greater than threshold concentrations.

MEASUREMENT OF ODOUR INTENSITY

The perceived odour intensity (I) of an odorant is related to its concentration (c) by Stevens' law [2], expressed in logarithmic form as:

$$\log I = \log I_t + n \log \frac{c}{t} \qquad (1)$$

where n is an intensity index, I_t is the intensity at threshold concentration and c/t is the concentration of the odorant in odour units.

The intensity of an odorant may be estimated by comparison with a butanol reference odour. For any odorant, equation 1 becomes:

$$\log I_o = \log I_t + n_o \log \frac{c_o}{t_o} \qquad (2)$$

and for the butanol standard:

$$\log I_s = \log I_t + n_s \log \frac{c_s}{t_s} \qquad (3)$$

where n_o and n_s are intensity indices for the odorant and butanol respectively. When the two intensities are equal:

$$\log \frac{c_s}{t_s} = \frac{n_o}{n_s} \log \frac{c_o}{t_o} \quad (4)$$

For each point of equal intensity, the concentration of the butanol standard may be plotted against the odorant concentration on a log-log basis. The slope of the line through the experimental points will give the intensity index ratio n_o/n_s. The intensity index for butanol as recommended by ASTM E544 is 0.66 [3] and hence the intensity index for any odorant may be found.

EXPERIMENTAL

The odorants chosen were ethyl acetate, iso-amyl acetate, 1,1-diethoxyethane (acetal), dimethyl trisulphide (DMTS), guaiacol, 2-nonenal, iso-valeraldehyde and vanillin. The threshold concentration of each odorant in a medium of 23%v ethanol was first estimated. Various dilutions were presented to panellists in nosing glasses in a triangle test, using 23%v. ethanol as the reference solution. The threshold concentration was taken as the lowest to be significant at a probability level of 0.01 [4].

Ten butanol solutions, each containing 50% more than the previous one, were prepared in 23%v. ethanol giving a set of reference intensities starting from ten times threshold level. A similar set of solutions of the odorant, also in 23%v. ethanol, was prepared and presented in nosing glasses. Panellists were asked which butanol reference odour was closest in intensity to that of the odorant. The average butanol concentration was plotted against the odorant concentration.

RESULTS

Figure 1 gives the intensity plot for acetal, with concentrations in mg/litre, and Figure 2 shows the intensity plots of all odorants, normalised by using odour unit concentrations (c/t).

Table 2 gives the thresholds and intensity indices for all the odorants. Each index was measured with a correlation coefficient between 0.96 and 1, statistically significant to a confidence level of 99.9%.

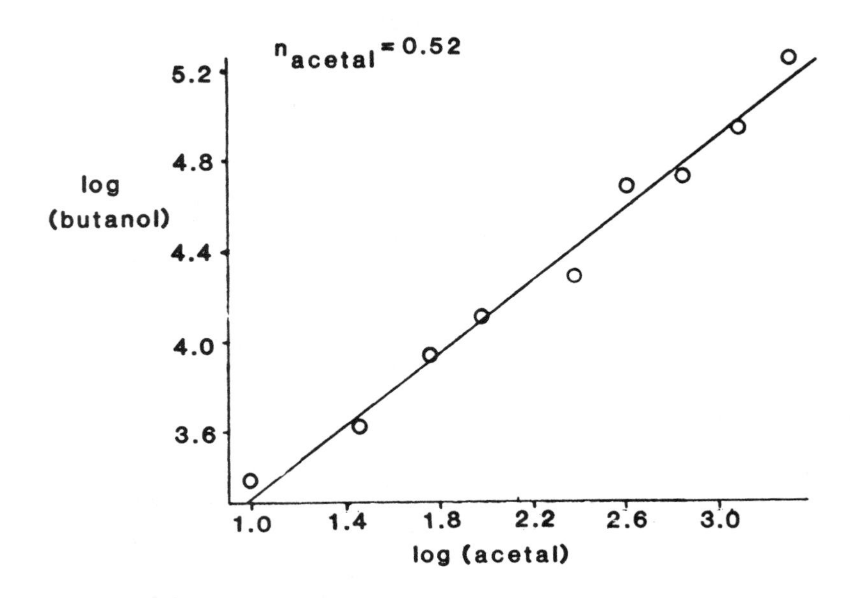

Figure 1. Intensity Estimation of Acetal

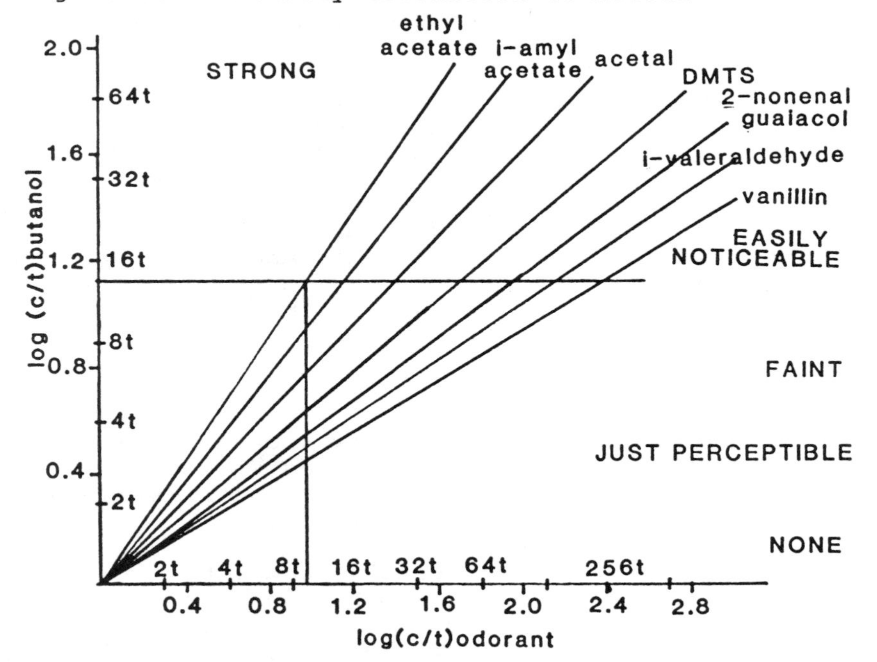

Figure 2. Odour Intensities

Table 2. Nosing parameters of aroma compounds (in 23%v. ethanol)
(o.u. = odour unit concentration $^c/t$)

Compound	Threshold (mg/litre)	Intensity Index	Concentration at equal intensity o.u.	mg/litre
ethyl acetate	74	0.74	10	740
iso-amyl acetate	1	0.63	15	15
acetal	1	0.52	27	27
DMTS	0.0001	0.43	54	0.005
guaiacol	0.06	0.38	96	6
2-nonenal	0.0006	0.38	96	0.06
i-valeraldehyde	0.05	0.34	153	7.7
vanillin	0.1	0.31	264	26

DISCUSSION

The esters possessed the largest intensity indices and the aldehydes the smallest, as observed by Jounela-Eriksson [5]. No relationship was found between threshold concentration and intensity index.

The variation of the intensity index explains why the odour contribution of each compound differs at the same odour unit concentration. On Figure 2 a line of equal intensity is drawn through a point where ethyl acetate is present at ten times its threshold concentration. The last two columns of Table 2 show the concentrations of the other compounds required to give the same intensity. Thus 27 units of acetal and 260 units of vanillin have the same intensity as 10 units of ethyl acetate.

Dravnieks [6] has described a category scaling method in which odour intensities are related to descriptors: (5 .. very strong, 4 .. strong, 3 .. easily noticeable, 2 .. faint, 1 .. just perceptible, 0 .. none). These descriptors have been added to Figure 2 so that odour strengths at various concentrations may be assessed. Thus, acetal in a pure medium would be just perceptible at 2t, and easily noticeable at 20t.

ODOUR CONTRIBUTION OF WHISKY COMPOUNDS

The odour contribution of the compounds to the total aroma of whisky may now be assessed. Table 3 gives typical concentrations of odour compounds in new and mature whiskies, and shows the category of odour intensity that would be observed in a single component solution. However, the perception of odours in mixtures may require higher concentrations.

Table 3. Odour contributions to malt whisky (after dilution to 23% ethanol)
M ... mature, N ... newly made

Compound		Concentration (mg/litre)	o.u.	Odour category
Ethyl acetate	(M)	150	2	just perceptible
iso-amyl acetate	(M)	6	6	faint
acetal	(M)	20	20	easily noticeable
DMTS	(N)	0.005	50	easily noticeable
guaiacol	(M)	0.1	1	none
2-nonenal		not known		
i-valeraldehyde	(N)	7	140	easily noticeable
vanillin	(M)	1	10	just perceptible

The presence of dimethyl trisulphide and iso-valeraldehyde in new spirit, and acetal in mature spirit is emphasised. Damascenone, present at a concentration of 2500 odour units in whisky, might be expected to dominate whisky aroma. The observation that the odour is not noticeable means that damascenone must have a very small intensity index.

CONCLUSION

The importance of an aroma compound may be determined by measurement of three parameters: the concentration of the compound in the sample, the threshold concentration and the intensity index. The odour intensity may be described according to the concentration present. The parameters may be used to estimate:

(1) the change in concentration required to make a significant change in odour intensity,

(2) the change in intensity that might result from a variation in the concentration of a particular congener.

ACKNOWLEDGEMENTS

Thanks are due to the Directors of United Distillers p.l.c. for permission to publish this paper, and to Dr. P.J. Hardy and Mr. J.H. Brown of this Company for their work on damascenone.

REFERENCES

1. Demole, E., Enggist, P., Sauberli, U., Stoll, M. and Kovate, E., Helvetica Chimica Acta **53**, 541, 1970.
2. Stevens, S.S., American Scientist **48**, 226, 1960.
3. ASTM E544, Am. Soc. Test. Materials, Philadelphia 1975.
4. Roessler, E.B., Pangborn, R.M., Sidel, J.L. and Stone, H., Journal of Food Science **43**, 940, 1978.
5. Jounela-Eriksson, P., Lebens-Wiss u-Technolog **15**, 303, 1982.
6. Dravnieks, A., American Chemical Society Symposium Series - Flavour Quality: Objective Measurement Symposium **51**, 11, 1977.

19

A peak matching technique for capillary GC data and its use in Scotch whisky flavour studies

Steven B. Muller, D. Howie and J. S. Swan
Pentlands Scotch Whisky Research Ltd., 84 Slateford Road,
Edinburgh, EH11 1QU, Scotland

The increased resolving power of capillary GC columns over packed columns places a greater strain on the peak identification properties of computing integrators. A peak matching technique based on multiple reference peaks has been developed and is used to enable multivariate analysis of the resulting data.

Identification of GC peaks by a single reference peak method is based on relative retention times. These are the ratio of the retention time of a component to that of a reference. A preset tolerance on the relative retention time is then used to provide a window for peak identification. With large numbers of closely eluting peaks these windows overlap and peaks can be wrongly identified.

The use of multiple reference peaks can overcome this problem. In 1958 Kovats proposed a method of retention indices based on the addition of a series of n-alkanes. Such mixtures were chromatographed isothermally and peaks identified consistently by reference to the added standards.

Isothermal GC is rarely used today in flavour research. Linear temperature programing is necessary to exploit the resolving power of modern capillary columns. Because of this a modified form of Kovats equation was developed. It was unnecessary to add a series of standards because most natural products, including whisky will contain homologous series of compounds which can be used e.g. fatty acid esters. A technique using a similar approach has been applied to tandem column GC analysis by the Finnish Customs Laboratory. (Kiviranta, 1987).

Integrated report files are processed by a suite of Basic programs. These calculate retention indices based on the known reference peaks, compare the retention indices with a data base of peaks of interest and check the matched data for missing peaks. The data can then be arranged into a form suitable for further analysis by multivariate techniques.

Using this approach sets of samples can be separated into subgroups on the basis of a particular attribute e.g. steam volatile phenols. The peaks responsible for the clustering of the samples into these subgroups can be identified and further sensory analysis carried out.

REFERENCES

Kovats, E., (1958), Helv.Chim.Acta 41, 1915.

Kivaranta, A., (1987), International Laboratory January/February, 58 - 65.

20

Flavour from speciality malts

T. M. Morris and C. W. Towner
Brewing Research Foundation, Lyttel Hall, Nutfield, Redhill,
Surrey, RH1 4HY, UK

ABSTRACT

Speciality malts are used in brewing to impart desired colour and flavour characteristics to beer. These malts are produced using prolonged roasting and kilning schedules which lead to the formation of higher concentrations of the flavour-active compounds typically present in "white" malt. Therefore, there is potential for using these speciality malts as a source of flavour by obtaining concentrated extracts which can be used to modify and enhance the malt flavour characteristics of beverages. Advantages of this approach include greater control over the consistency of flavour, increased utilisation of flavour-active components and flexibility in adjusting the flavour of the final product.

Initial investigations have concentrated upon identifying the major desirable flavour notes derivable from these malts and relating these to the group or groups of compounds responsible. Ultrafiltration has proved useful for the separation of compounds of different molecular weights and allows extracts to be prepared which can impart significant flavour effects without the need for an accompanying increase in colour.

INTRODUCTION

Speciality malts (eg Crystal Malt, Carapils, Chocolate Malt and Roasted Barley) are widely used throughout the Brewing Industry to impart both colour and flavour characteristics to beer. As these malts also provide a significant contribution as a source of extract, the flavour-active components within them are subjected to the rigours of the whole of the brewing process. The majority of these compounds remain chemically unchanged throughout the various processing stages. Although more of these compounds are formed during wort boiling, overall significant physical losses occur with the result that only a small percentage survives into the final beer. In addition, the degree of loss can vary from brew to brew leading to inconsistencies in the final flavour.

In brewing, a similar situation exists with regard to traditional hopping protocols and recent advances (Westwood, 1985) have been made in isolating desired flavour components from hops in forms which can be added directly to the final beer. This approach has the advantages of providing greater control over the consistency of flavour, increased utilisation of flavour-active materials and flexibility in adjusting the flavour of the final product. Clearly, there is scope for applying this same approach to malt-derived flavours which would be applicable not only to beer but to all beverages which rely upon malt-derived flavours as a desirable, intrinsic part of their character.

Speciality malts have, therefore, been investigated as potential sources of flavour concentrates.

NATURE OF SPECIALITY MALTS

Speciality malts are produced by roasting and kilning at higher temperatures than those employed to produce standard malted barley. The degree of kilning, the nature of the malt, barley or other cereal prior to kilning governs the nature of the final product. A range of products exists. Typical, basic properties of some of these are shown in Table 1.

Crystal malt, which with carapils accounts for about 75% of speciality malts made in the UK, is a sweet, caramel flavoured malt which has undergone enzymic conversion of its starch into sugar during the early stages of roasting. This is achieved by heating the malt in a high moisture environment which produces a "stewing" effect.

This stage takes approximately 40 mins. with the temperature rising to about 100°C before full conversion is achieved. Moisture is then removed and the temperature increased to about 135°C to produce the desired colour.

TABLE 1 - Properties of Speciality Malts

Speciality Malt	Extract (dry) (Litre° kg^{-1})	Moisture (% w/w)	Colour (°EBC)
Standard Ale Malt	308	3.7	8
Standard Lager Malt	309	4.9	3
Crystal Malt	273	4.2	140
Carapils	275	6.7	27
Brown Malt	270	2.3	130
Amber Malt	277	<2.0	55
Chocolate Malt	267	<2.0	1100
Roasted Barley	270	<2.0	1400

Initial studies have concentrated on the flavours arising from the use of crystal malt which have been variously described as malty, nutty, caramel, burnt, toffee-like, biscuity, roasted and astringent.

ANALYSIS OF FLAVOUR COMPONENTS

Analysis of the compounds responsible for the distinctive flavour characteristics of Speciality Malts has been carried out using Gas Chromatography coupled with Mass Spectrometry (GC/MS). Most attention has focussed on both oxygen and nitrogen containing heterocylic compounds which are well known products of non-enzymic browning or Maillard reactions. These reactions occur between sugars and amino acids and products from them have been shown to be responsible for the flavour of many different foodstuffs (Maga, 1973). Many such compounds have previously been identified in beers brewed using highly coloured malts (Tressl, 1977).

Methods of analysis have been developed which can be applied equally well to malts, worts or beers. In the case of malt, finely ground malt (250g) was stirred in water (1000ml) for one hour at ambient temperature. The resulting suspension was filtered. The filtrate was treated in the same manner as for a wort or a beer. Internal standards were added as required. Ultrafiltration was carried out across a membrane which retains compounds of >100,000 Nominal Molecular Weight. The permeate (1000ml) was continuously extracted with dichloromethane for 16 hours. The organic phase was separated, dried and the solvent evaporated under reduced pressure. The residue was redissolved in ethanol (50ml), re-evaporated under reduced pressure and the final volume adjusted with ethanol to 10ml. An aliquot (1-2μl) was injected into a Finnegan 1020 GC/MS instrument containing a 25m x 0.32mm i.d. WCOT 1.2μ df CPWax 52CB fused silica capillary column using a splitless injection technique. Helium was used as the carrier gas (linear velocity 35cm/sec). The GC temperature programme ran from 60°C (3 min) to 200°C at 3°/min. Selective Ion Monitoring (SIM) was used to detect and identify the individual compounds of interest.

Using this method many compounds have been detected in beers brewed using crystal malt and in extracts of crystal malt. A number of these have been identified as potential flavour contributors and used as marker compounds for assessing the efficiency of extraction and for attempting to relate analytical values to perceived flavour characteristics. These selected compounds are listed in Table 2 along with their measured levels in an extract of crystal malt.

TABLE 2 - Levels of Marker Compounds in an Aqueous Extract of Crystal Malt

Compound	Concentration (ppb)
2-methylpyridine	0.5
2-methylpyrazine	173
2,5-dimethylpyrazine	7
2,6-dimethylpyrazine	5
2-ethylpyrazine	2
2-ethyl-6-methylpyrazine	2
2-ethyl-5-methylpyrazine	1
Trimethylpyrazine	3
isomaltol	nq
cyclotene	145
maltol	9000
5-hydroxy-5,6-dihydromaltol	3000

NOTE nq = detected but not quantified

FLAVOUR EFFECTS

The effects of an extract of crystal malt on the flavour profile of a beer is illustrated in Figure 1. The flavour descriptors used cover the majority of flavour characteristics attributed to Speciality Malts and form the basis of a specialised profile format which has been of particular use for comparing beers and extracts prepared from different malts using different procedures.

A straightforward aqueous extract of crystal malt is illustrated which has passed through an ultrafiltration membrane (100,000 NMW). Clearly, the flavour effects from adding this extract to beer are similar to those produced from incorporating crystal malt in the grist. However, to achieve a similar intensity of flavour, only approximately one tenth of the amount of crystal malt need be extracted. Some adjustments can be made to the flavour effect produced either by modifying the conditions of extraction or carrying out further extraction and fractionation procedures. In particular, the ratio of burnt, astringent notes to malty, toffee notes can be varied. Analysis has shown that this correlates with the ratio of nitrogen-containing to oxygen-containing hetrocyclic compounds. It is felt that the flavours resulting from the latter group of compounds are more desirable.

FIGURE 1 Flavour Profile Assessment of an Aqueous Extract of Crystal Malt

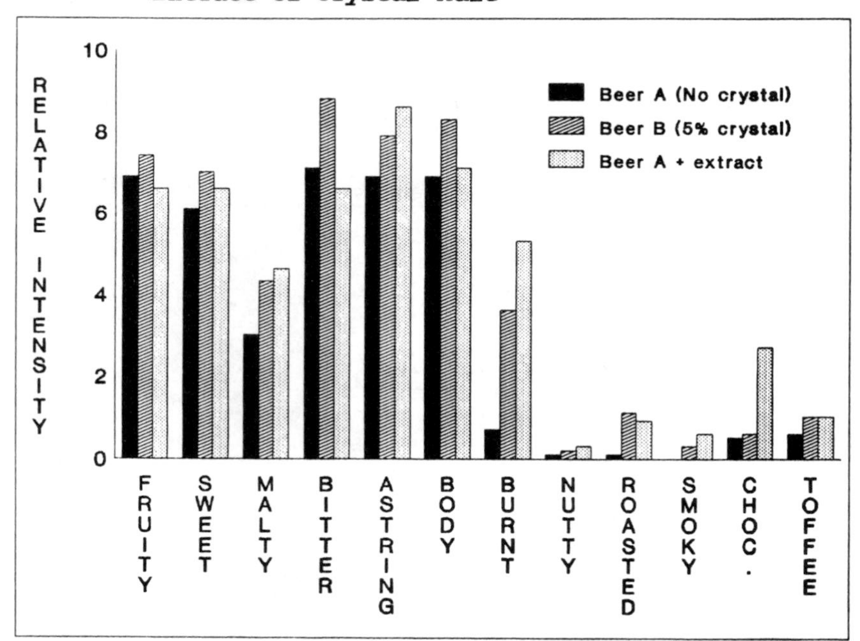

EXTRACTION PROCEDURES

A number of options exist for the extraction and isolation of flavour components from Speciality Malts. These tend to fall into the two categories illustrated in Figure 2. The simplest method is the direct extraction of milled malt with an appropriate solvent and a number of solvents have been investigated ranging from liquid carbon dioxide to methylene chloride. The selectivity of different solvents for different compounds causes some variation in flavour effect but options for controlling the selectivity of extraction for each individual solvent are somewhat limited. Consequently, although extraction efficiencies are high, some undesirable material is also extracted. Lipid material is particularly deleterious as this imparts an unpleasant fatty, oily note to the beer. Further processing would be necessary, therefore, in order to remove such undesirable contaminants. Nevertheless, the fact that sugars would not be extracted but would remain available for use in fermentation and also that benign solvents such as ethanol or carbon dioxide could be used, makes this approach reasonably attractive.

FIGURE 2 Options for the Extraction of Flavour Compounds

1. Solvent extraction of milled malt using :
   - Liquid Carbon Dioxide
   - Supercritical Carbon Dioxide
   - Organic Solvents
     e.g. Ethanol, Hexane,
     Methylene Chloride.

2. Aqueous extraction of milled malt followed by :
   - Ultrafiltration
   - Freeze Concentration
   - Solvent Extraction (as above).

The alternative approach is very similar to that used for the analysis of the flavour components and involves an initial extraction of milled malt with water. In this case, lipid material is not extracted although, dependent on the type of malt used, some of the sugars are. Volume considerations dictate that a further concentration stage is required to provide a viable product. However, in achieving a more concentrated extract it has been found possible to introduce some selectivity and control over the nature of the compounds present in the final product. As the majority of the flavour-active fraction is comprised of relatively small molecules, the process of ultrafiltration is particularly suited to removing undesirable material of large molecular weight. Use of membranes which retain material of >100,000 NMW removes proteinaceous material and allows any subsequent solvent extraction to be performed without problems of foaming or emulsion formation. Much of the colour can be removed by membranes which retain material of >10,000 NMW. This provides a mechanism for separating flavour from colour and can result in a flavoursome extract of relatively low colour, whilst the highly coloured fraction may be used for adjusting colour without a significant, accompanying flavour effect. Concentration of the flavour components can be achieved using freeze concentration techniques or by solvent extraction of the aqueous solution. Adjustment of pH prior to solvent extraction can be used to favour the extraction of either acidic (eg pyrones) or basic (eg pyrazines) compounds and thus to modify the resulting flavour.

The optimum extraction procedure depends, therefore, both on the economic factors relating to each processing stage and on the desired nature of the extract.

CONCLUSION

Initial investigations have shown that Speciality Malts provide a useful source of desirable flavour components which can impart pronounced effects on the flavour of beer.

Use of judicious processing in the extraction and isolation of these components can produce a range of flavoursome extracts of differing character which need not necessarily be also high in colour. Progress has been made towards characterising these extracts both from analytical and flavour aspects. Further work is required to optimise the procedures for producing such extracts but the potential of these materials for expanding the range of flavours present in malt-based beverages has clearly been demonstrated.

REFERENCES

Maga,J.A. and Sizer,C.W., CRC Crit.Rev.Fd.Technol.,39, (1973)

Tressl,R., Renner,R., Kossa,T. and Koeppler,H., Proc.Eur. Brew.Conv.(Amsterdam),693,(1977)

Westwood,K.T. and Daoud,I.S., Proc.Eur.Brew.Conv. (Helsinki),579,(1985)

21

Opportunities for product development

Michelle Proud
County NatWest WoodMac, 74-77 Queen Street, Edinburgh, EH2 4NS, Scotland

Important opportunities for product development in spirits will open up within Europe over the next few years. I shall focus on Europe because two key factors that help create such opportunities - market growth and changing patterns of consumption - will be particularly evident in Europe over the next few years. Part of this is happening anyway, but a major stimulus will come through the changes that will take place as part of the process of turning the EEC into a single internal market from the end of 1992.

First, I will look briefly at the market as it is and at what opportunities emerge as a result of current patterns of consumption. Then I will consider what changes are likely to take place between now and the early 1990s and what the impact of the changes will be. Finally, I will look in more detail at the types of product development likely to take place in spirits.

THE EUROPEAN DRINKS MARKET - WAITING TO BECOME ONE

The 12 EEC countries, with a population of 323 million and a combined GNP of $4226bn, represent potentially the largest economic group in the non-communist world. But in performance they are surpassed by both Japan and the US.

The figures in Table 1 illustrate the way that the EEC lags behind its major competitors in terms of wealth creation. It is this kind of discrepancy that the creation of a single market under the "1992" plan is intended to combat. The figures show the scope for growth in consumer spending - which will include growth in spending on alcoholic drinks - as the 12 countries of the EEC catch up, with each other and with the US and Japan.

Table 1 : Population, GNP and Consumer Spending, 1987

	Population 10^6	GNP ECUx10^9	GNP Per Head ECU '000	Consumer Spending/Head ECU '000
Germany	61.1	963.6	15.8	10.0
Denmark	5.1	87.7	17.2	9.3
France	55.6	748.7	13.4	8.2
Luxembourg	0.4	5.3	14.3	8.1
Belgium	9.9	120.1	12.2	7.7
Netherlands	14.6	183.5	12.6	7.5
Italy	57.3	645.5	11.3	6.9
UK	56.8	565.6	10.0	6.2
Spain	38.7	245.9	6.3	4.1
Eire	3.6	24.7	6.9	4.0
Greece	10.0	40.6	4.1	2.7
Portugal	10.3	30.5	2.9	2.0
Total EEC 12	323.4	3661.4	11.4	7.0
US	244.0	3838.4	15.7	10.5
Japan	122.4	2018.2	16.5	9.6

TRANSCENDING THE BUREAUCRATIC RESTRAINTS

The first question is what "1992" means and the chances of it actually happening. The specific definition is that "1992" is the plan, embodied in some 300 detailed directives, whereby physical frontiers between EEC countries will be abolished on 31st December 1992.

At this stage, it seems safe to say that all 300 directives will not be adopted in time. It is almost equally certain that not all will be found to be necessary, but that new problems will emerge that will have to be dealt with.

How many directives are approved, and by what date, is not, however, the main point. **1992 is not a date; it is an idea. More important - it is a state of mind.** It is beginning to happen already and its implications will have an accelerating impact, because companies and governments are starting to plan on the assumption that it will happen.

For the drinks industry, 1992 will see an intensification of processes already happening elsewhere in the world, particularly in the US, and already evident in Europe. These are :

- o increasing concentration in production facilities;
- o increasing market share for leading pan-European/international brands;
- o the development of EEC-wide distribution systems by the major groups, through acquisition and joint ventures;
- o modification of the traditional system of exclusive national for brand distribution, leading eventually to replacement by other systems.
- o a gradual convergence in patterns of consumption throughout the EEC. This will occur in the level of consumer spending on drink and the level of alcohol consumption; the balance between the three major categories, beer, wine and spirits; and the choice of brand and product within these categories.

A FRAGMENTED MARKETPLACE...

The European drinks market is enormously fragmented. In spirits, the dominant, locally-produced product is different in most of the 12 EEC countries. Local availability of raw materials has been one of the chief determinants, but so have accidents of trade. The long history and level of sophistication in local production has made such differences far more deeply entrenched than in "newer" markets like the US.

Inevitably, the level of brand fragmentation has reflected this. Only in drinks categories which have been specifically developed as exports by large companies - such as Scotch whisky from the UK or vermouth from Italy - are there measurable pan-European brand shares.

...BUT CONVERGENCE TRENDS ARE IDENTIFIABLE

Already this is changing fast. The parallel trends of convergence in consumption and brand concentration are producing a more homogeneous pattern of drinking throughout the EEC. Beer consumption is rising fast in the traditional wine-producing countries of the south, notably in Italy, while wine consumption is growing fast in traditional beer-drinking countries of the north, notably in the UK. Throughout the EEC, the international spirits categories such as Scotch, gin and vodka, are gaining share in new markets.

The first opportunity that might be expected to emerge in the EEC would be in countries with a below-averge level of alcohol consumption. Average per capita consumption in the EEC is just over 10 litres pure alcohol per year. In most of the biggest markets - France, Germany, Spain and Italy - consumption is already above this level, and the trend has been downwards. There is, moreover, no obvious correlation between alcohol consumption and spending power - Portugal ranks second among EEC countries in volume of alcohol consumed, though it ranks bottom among the 12 in the level of consumer spending (the local wine is obviously very cheap).

Table 2 : Alcohol Consumption Per Head in EEC, 1986

(litres pure alcohol)	Spirits	Wine	Beer	Total
France	2.1	9.6	2.0	13.7
Portugal	1.0	9.7	1.9	12.6
Germany	2.5	3.0	7.0	12.5
Luxembourg	1.6	4.5	5.8	11.9
Spain	3.2	5.4	3.2	11.8
Italy	1.4	8.4	1.1	10.9
EEC Average	1.9	4.5	3.9	10.4
Denmark	1.6	2.2	6.2	10.0
Belgium	1.3	2.7	5.7	9.7
Greece	2.7	5.1	1.4	9.2
Netherlands	2.2	2.1	4.1	8.4
UK	1.7	1.7	4.0	7.4
Eire	1.7	0.5	4.7	6.9

The UK is the one substantial market with distinctly below-average consumption, at only 7.4 litres pure alcohol per head. Consumption has been constrained in recent decades by the high level of excise duties, and might be expected to rise as and when excise duties are lowered. However, social factors - ranging from health consciousness to anti-drink-driving legislation - are likely to mean that when the spending boom for alcohol occurs, it will take the form of better quality, not higher quantity. The opportunity here is clearly for premium brands - as it also is in France and Germany as volume consumption declines.

The convergence theory of consumption assumes a gradual shift away from traditional preferences, towards international categories, on a rough split (in terms of pure alcohol) of 40% each for beer and wine and 20% for spirits. Using this as a yardstick, a clearer set of opportunities appears, since nowhere in the EEC is this pattern yet evident. The growth markets on this basis are:

- o for beer : Spain, Italy, Portugal and Greece;
- o for wine : Germany, Belgium, Denmark, the UK and Eire;
- o for spirits : all countries apart from France, Germany, Spain and Greece.

The fact that three of the last-named countries are currently particularly fast-growng markets for Scotch whisky appears to disprove this. What is happening, however, is that newer imported categories are replacing traditional domestic spirits. This is where the biggest opportunities lie.

[MPORTS v DOMESTIC

As an illustration, in two of the larger spirit markets - France and Germany - nearly half of the spirits consumption is accounted for by products which are not only domestic but are also little drunk elsewhere, such as korn in Germany and pastis in France. Likewise in both, roughly a quarter of consumption is accounted for by imported products - imported predominantly from elsewhere in the EEC. After 1992, these will no longer be "imported" products: they will be fashionable, prestigious, international brands competing equally on price with domestic spirits of like quality.

Table 3. EEC Market For Alcoholic Drinks, 1986

	Spirits: $x10^6$ 9-litre cases	Wine cases $x10^6$	Beer litres $x10^6$
France	33.4	494.2	2126
Germany	42.5	169.8	8908
Italy	24.8	46.5	1310
Spain	36.6	193.6	2382
UK	27.9	107.9	6121
Netherlands	10.5	26.0	1226
Denmark	2.2	11.3	658
Belgium	3.9	24.2	1182
Greece	7.6	48.8	320
Portugal	3.0	96.1	380
Eire	1.7	1.6	432
Luxembourg	0.2	1.8	48
EEC Total	195.0	1221.6	25093

Sources : CNWM estimates

Volume and Margin Growth Prospects for Alcohol

	Consumption Growth Prospects			
	Beer	Wine	Spirits	Total Alcohol
UK	-	###	#	###
Denmark	●	##	#	#
Netherlands	-	###	●	##
Eire	●	###	#	###
Belgium	●	##	##	#
France	##	●	●	●
Greece	###	●	●	##
Italy	###	●	##	-
Portugal	##	●	##	●
Spain	#	●	●	●
Germany	●	#	-	●
Luxembourg	●	-	#	●

	Impact on Volume of Tax Changes			
	Excise			VAT
	Beer	Wine	Spirits	All Alcohol
UK	###	###	###	-
Denmark	##	###	###	#
Netherlands	-	-	-	-
Eire	###	###	###	##
Belgium	##	-	-	#
France	●	●	●	-
Greece	●	●	●	#
Italy	-	●	●	-
Portugal	●	●	●	-
Spain	●	●	●	-
Germany	##	●	●	-
Luxembourg	●	●	●	-

	Overall Growth Prospects	
	Volume	Margin
UK	#	###
Denmark	##	###
Netherlands	##	##
Eire	#	##
Belgium	##	#
France	●	#
Greece	-	#
Italy	##	●
Portugal	#	●
Spain	●	●
Germany	●	●
Luxembourg	-	●

Growth ●Decline

THE CHANGES AHEAD OF 1992

1992 is going to happen through two parallel processes: the removal of existing barriers to create a single "internal" market and the strategy of companies in treating the whole EEC as a single market.

The first is only designed to facilitate the second. However, removing barriers is a misleading description: the process is actually one of replacing 12 existing and varied sets of regulations with a single common set of regulations. The main aspect of these changes to affect the drinks industry is the harmonisation of excise duties and VAT, but changes will be taking place in related areas such as packaging (bottle sizes etc.), labelling, product specifications, transport, advertising, licensing laws and competition laws. This process is itself so complex that the impetus towards 1992 threatens to get bogged down in detail.

However, as the idea of 1992 gathers momentum - at government, corporate and consumer level - pressure will grow to keep the process as simple as possible. Some of the proposals now under review - particularly those on fiscal harmonisation - appear to require more changes than is really necessary. Compromises along the way will mean a slimming down in the number and complexity of changes ahead of 1992.

HARMONISING EXCISE DUTIES

The biggest stumbling block to free trade in alcoholic drinks in the EEC is the huge differential between the excise duty rates in different countries. A bottle of Scotch currently costing £7.50 in the UK would (all other things, including bottle size, being assumed equal) cost £9.44 in Denmark, £4.96 in France, £3.36 in Spain and only £2.86 in Greece, purely on the basis of the different duty rates. Using the proposed EEC rate, it would cost £5.19 throughout the EEC (Table 4).

THE NORTH-SOUTH DIVIDE

For countries with low rates - the four most southerly, Spain, Portugal, Italy and Greece - the governments might rejoice at the additional revenues they might get from such a big increase in excise rates, but consumption would be devastated. These four also have at the moment no duty at all on wine, and low rates on beer (except for Italy, which charges more on beer than most "medium rate" countries).

For countries with high rates - Denmark, Eire and the UK - government revenues would be cut severely, though not to the full extent of the change in duty rates, as consumption would be likely to rise (to the dismay of the anti-alcohol lobby).

Table 4. Excise Duty Rates in EEC Countries*

ECUS	Spirits per 70cl bottle	Still wine per litre	Beer per litre
High Rate Countries			
Denmark	9.80	1.57	0.71
Eire	7.63	2.79	1.13
UK	6.95	1.54	0.68
Medium Rate Countries			
Netherlands	3.63	0.34	0.23
Belgium	3.51	0.33	0.13
Germany	3.29	-	0.07
France	3.22	0.03	0.03
Luxembourg	2.36	0.13	0.06
Low Rate Countries			
Spain	0.87	-	0.03
Portugal	0.69	-	0.09
Italy	0.64	-	0.17
Greece	0.13	-	0.10
Proposed EEC Rate	**3.56**	**0.17**	**0.17**

*as at 1.4.86

The five remaining medium-rate countries would see comparatively little change, though there would be changes in the relative price levels, particularly of beer and wine. In the Netherlands, Belgium and Germany, these would favour wine; in Luxembourg they would favour beer. In France, which taxes beer and wine equally for a like volume, the impact would be to make both more expensive in relation to spirits.

WHERE IT GETS COMPLICATED

The proposed spirits rate, a straightforward tax on alcohol content of 1271.14 Ecu per hectolitre of pure alcohol, was arrived at simply by taking the arithmetic mean of the rates in force in the 12 countries (as they were, converted in Ecus, on 1st April 1986). This method could not be used for beer and wine because the result failed to preserve any sensible relationship between the two rates, and more arcane calculations were therefore used. This produced a final rate that works out at 17 Ecus per hectolitre for both light table wine and for beer of average strength.

Moving Into Balance : Excise Duties on Beer, Wine, Spirits

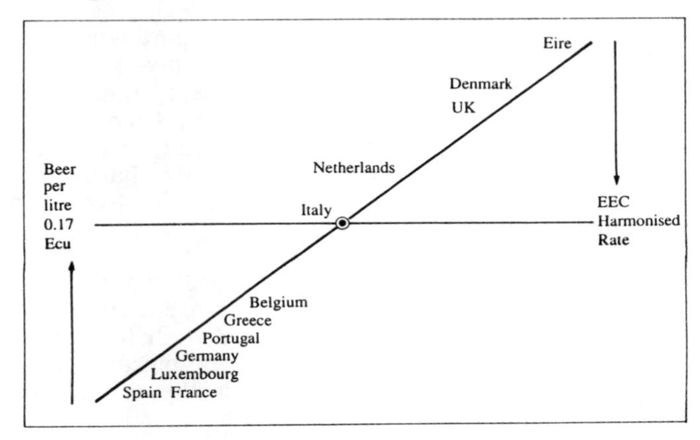

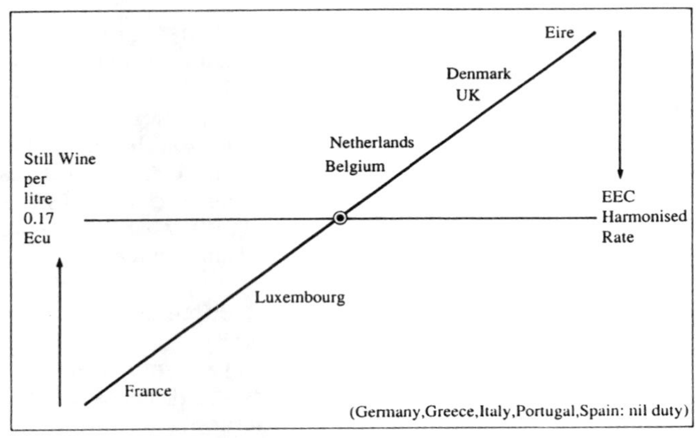

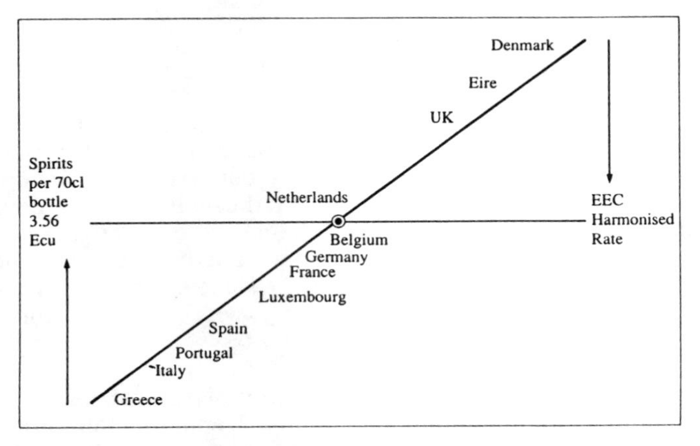

MUCH HORSE-TRADING IN PROSPECT

In setting these excise duty rates, the Commission was grappling with two contradictory aims: to remove all barriers to free competition between products while causing the minimum disruption - to fiscal revenues and to industry. Clearly, compromise is essential. Compromise has already been made, for instance, in proposong to tax beer more heavily than wine, and spirits far more than either wine or beer (which already happens in all 12 countries).

Further compromise will be needed. Indeed, the Commission recognises that its proposed rates represent a starting point for negotiation. The proposals are being studied both by the governments of each country and by trade organisations, including the Union Europeenne des Alcohols, Eaux-de-Vie et Spiritueux, which represents European spirits producers.

It is clear at this stage that both the timetable and the proposed rates represent too ambitious a target. One proposal, contained in a report by the Institute of Fiscal Studies, is that the EEC should be divided into three zones, each with different excise rates, but each moving towards common levels. The three high-rate countries - Denmark, Eire and the UK - are those which are furthest separated, geographically, from the four low-rate countries, which would minimise the possibility of cross-border trading to take advantage of the differential in rates.

Such zoning would represent a first step in an extended transitional process, with phased changes towards a common set of rates. This would also allow the possibility of more fundamental reform in the taxation of alcohol, to bring all rates on to a common basis, i.e. in relation to alcohol content.

GROWTH MEANS NEW OPPORTUNITIES

Whatever the outcome on the detail and timing of excise duty changes, the EEC proposals do represent one certainty: the direction of change. Duties in high-tax countries will be coming down - perhaps very gradually - while those in low-tax countries will be rising; and though spirits will continue to be taxed more heavily than wine or beer for the same alcohol content, the relative differential will gradually be reduced. Price movements will be favourable to spirits, producing greater growth for spirits relative to other alcohol products.

The prospect therefore is one of growth for spirits. In the northern countries, particularly the UK, Denmark and Eire, this is likely to mean some volume growth, but growth in spending terms will be more likely everywhere. What will this mean for new product development?

1. Consumers will move up-market to "international" categories. It is important to stress here that what constitutes an international category is not to do with product type but with historical trading developments. The British turned two domestic products - Scotch whisky and London-style dry gin - into international products by virtue of exporting them; the French did the same with cognac and liqueurs but failed, or neglected, to do so with their aniseed-based spirits.

Opportunities for product development in terms of existing "international" products are considerable - and probably, in the longer term, more profitable than developing new products. Although Scotch, for instance, has had a prestigious reputation in Southern Europe for many decades, its recent growth in countries such as Greece, Spain and Portugal has been essentially that of a new product (which has become available on a "popular" scale in these markets only since they joined the EEC). Gin is a key product in Spain but, along with Western-style vodka, is little drunk elsewhere in Europe.

2. "Domestic" categories, which are viewed as downmarket in their country of origin, still constitute around a third of the EEC spirits market, or around 65 million cases. As consumers switch to international categories, these categories will go into decline. They are, however, traditional spirits with a long pedigree of development; they represent a rich store of source material for development, in a process whereby the local specialities of one region can become the new products of the expanded European market.

In this context, domestic spirits include products such as grappa/marc, fruit brandies and liqueurs of all types, with considerable local variations, the bitter aperitifs of Italy and the herb/bitter liqueurs of both Italy and Germany. The list goes on and on: there is little need to create new products when so many products, known only in a limited region already exist.

There is one key point about all these products: they are "adult" spirits, mostly with a characteristic dry flavour: they are preferred by mature drinkers and are not suitable for consumers in their 20s, for whom the majority of new spirits products have been developed. Demographic changes mean that older age groups - people in their 40s, 50s, 60s and over - will become increasingly important as consumers, by vitue of the growing proportion of the total population that they will represent in both numbers and affluence. Yet they have been neglected in terms of product development and marketing.

Creating modern new products for this target market out of the traditional spirits of different localities is primarily a marketing job. But it should be easier to do with a product with some tradition behind it, and hence a story to tell, than with an "invented" product. A final advantage to such products, apart from the taste profile, is the fact that they are "genuine" local products. Invented new products are often regarded in some way as ersatz - fun to try, but not a serious product to stay with, and not one which will contribute much to the status of the drinker.

3. There are, finally, the "genuine" new products - or, to describe them more accurately, the genuinely novel combinations. Almost all new products are essentially that: a well-known spirit base, such as Scotch, white or dark rum, or other spirit, (immature Scotch being one of the less known) combined with known flavour or flavours, with perhaps another element to give it a novel consistency (dairy cream), colour/opacity or texture. In the handful of major success stories - Baileys Irish cream and Malibu in the UK (and now, increasingly, in Europe), De Kuyper Peachtree in the US, Berentzen Appel in Germany - the specific details have all been important contributors to the brand's personality.

New products like these have come to hold significant market shares in spirits in a number of countries. In the UK, they represent roughly 40% of speciality spirits and 4-5% of total spirits; in France 17% and 2% respectively, and in Germany 26% and 5%.

Companies will continue to invest in research and development of new products, but the pattern of this will be somewhat different in the future from the past. One reason for this is that the next few years will see a rapid process of concentration at company level in production, brand ownership and distribution. The highly fragmented brand shares across Europe in spirits will gradually be consolidated through merger and acquisition. The main groups which emerge through this process will be working on developing and fitting together brands in existing categories, and taking those brands into new markets; this will become a higher priority than the development of new brands from scratch.

A second factor will be a change in the marketing focus. The younger drinker, usually aged under 30, has been the target market for the bulk of new product development in the past few years. This age group is becoming steadily less important within the market. The emphasis in new product development is likely to shift from fashionable, sweet-tasting products geared to this age group to developing products with wider appeal. A key challenge, for instance, is the production of low alcohol products which do not just taste like watered-down versions of full-strength products.

The process of developing a spirits-based drink which both attracts consumers and can be taken seriously has become much harder - not only because of the few big successes, but because of the many me-too products. despite this, none of the major groups can risk missing out on an opportunity by failing to maintain a line new product development unit. But the emphasis will go on developing opportunities for existing brands and existing products, major or minor.

22

Marketing aspects of new product development

M. J. Baker
Department of Marketing, University of Strathclyde,
Glasgow, G4 0RQ, Scotland

THE IMPORTANCE OF NEW PRODUCT DEVELOPMENT

While it is true that demand does exceed supply in certain markets, this is largely confined to new markets for new products. In more traditional and familiar markets, such as those for steel, ship-building, car tyres and even cars themselves, potential capacity far exceeds effective demand, with the result that firms compete with each other for the customer's favour. This is particularly true of the market for fast moving consumer goods (fmcg) such as food stuffs, household cleaning materials, etc. Classical economics would have us believe that this competition would focus on price, but the reality is that manufacturers choose to recognise that demand is not homogeneous and so seek to differentiate their product to match more closely the needs of specific subgroups or market segments. Clearly, if the manufacturer is successful in distinguishing his product in a meaningful way, then he provides prospective users with a basis for preferring it over other competing alternatives and so creates a temporary monopoly which allows him a measure of control over his marketing strategy.

The desirability of exercising some control over a

market rather than being controlled by it - being a price maker rather than a price taker - is self evident. The wisdom of doing so is equally compelling, for a number of surveys have shown conclusively that while price ranks third or fourth in the selection criteria of most buyers, product characteristics or 'fitness for purpose' invariably ranks first.

For these reasons, product differentiation has become the basis for competition between suppliers competing for a share in a market and this explains the importance attached to product development by most firms today. It also helps to explain why new product development (or innovation) has become increasingly sophisticated and much riskier than it used to be. This is because, with so many more firms investing heavily, minor features are quickly copied or made obsolete by more radical changes. In the same way, the accelerating rate of change has had a similar effect on major innovations so that the average life of products is becoming shorter and shorter (compare, for example, valves, transistors and microprocessors as basic inputs to computing devices). Thus, many firms are faced with the apparent paradox that if they do not innovate they will be left behind, while conversely if they do the probability of failure is very high and this could ruin the company too. As Philip Kotler has put it:

"Under modern conditions of competition, it is becoming increasingly risky not to innovate ... At the same time, it is extremely expensive and risky to innovate. The main reasons are: (1) Most product ideas which go into product development never reach the market; (2) Many of the products that reach the market are not successful; and (3) Successful products tend to have a shorter life than new products once had."

With regard to the first point, it is clear that discarding products during development must impose some cost. However, there is a great deal of advice on this phase of development, and it is believed that companies have become much better at weeding out weak ideas earlier in the development cycle and so minimise losses from this source. Similarly, a shorter life might be preferable to a long one if one can generate similar volumes of sales, because the discounted value of present sales is greater than future ones and early capitalisation of an investment gives the firm greater opportunity for flexible action. This point will be developed later.

Much the most important cause for concern is the fact that many products are not successful at all. In these cases not only has a company incurred all the development costs, but it has also incurred the marketing costs of launching the new product, not to mention the possible loss of goodwill on the part of the users who discover the product is unsuccessful and likely to be withdrawn from the market.

THE NATURE AND CAUSES OF PRODUCT FAILURE

While claims concerning the incidence of new product failure are commonplace, few such claims are based on hard evidence. Those that are usually conflict with one another, due to the absence of any agreement about precisely what is to be measured, so that trying to quantify the proportion or value of failures is largely a matter of speculation. However, managers are agreed that the number and cost of failures is high and are anxious for advice as to how they can reduce this risk. In order to do so, it will be helpful to propose a simple definition of failure and then see if at least the major causes of it can be identified.

A simple definition of failure is that this is deemed to have occurred when the innovator so decides. While this may not appear to be very helpful, it should help to clear the ground by making it explicit that success and failure are comparative states and there is no yardstick or criterion for deciding when one ends and the other begins. To argue otherwise would be to claim that all firms subscribe to the same managerial objectives - for example, a return of x per cent on capital employed - and clearly they do not. It follows that your failure might be someone else's success and attempting to define the states precisely is a sterile exercise.

This is certainly not true of establishing the perceived causes of failure because, by doing so, it should be possible to develop guide-lines and tests for identifying and avoiding these in future. Unfortunately, relatively few firms appear to be willing to document their failures and there is a marked dearth of case history material on the subject. Many years ago (1964) the National Industrial Conference Board in the United States conducted a survey as a result of which it offered the following list of factors underlying failure in rank order of importance:

(a) inadequate market analysis;
(b) product defects;
(c) higher costs than anticipated;
(d) poor timing;
(e) competitive reaction;
(f) insufficient marketing effort;
(g) inadequate sales force; and
(h) inadequate distribution.

Over 50 per cent of all respondents cited the first three reasons.

A more recent study was carried out by Roger Calantone and Robert Cooper of McGill University, in which they asked managers in 150 industrial companies in Quebec to categorise the nature of the causes leading to market failure: 'those products where sales had failed far short of expectations'. Table 1 summarises the responses to this survey and reveals strong support for the findings of the 1964 study.

In between these studies, Andrew Robertson and his colleagues at the Science Policy Research Unit at Sussex University conducted an analysis of a series of 34 new product failures, and concluded that their main cause was a lack of market orientation.

While the evidence may not be as extensive as one might wish, the conclusion appears inescapable - failure is the consequence of managerial ignorance or, worse still, managerial neglect. Ignorance because there is a very extensive managerial literature, based on well documented practice, which emphasises the importance of thorough market analysis as an essential prerequisite of any new product development; neglect because it is management's responsibility to keep itself informed of the best current practice and, if one is well informed, it is difficult to conceive how one could excuse commercial failure in a variety of ways which fundamentally all comes down to the same thing - inadequate market analysis.

Table 1. Specific Causes for Poor Sales Performance (N = 89)

Specific Cause	% of product failures	
	Main Cause	Contributing Cause
Competitors were more firmly entrenched in the market than expected	36.4	13.6
The number of potential users was overestimated	20.5	30.7
The price was set higher than customers would pay	18.2	33.3
The product had design, technical or manufacturing deficiencies/ difficulties	20.5	25.0
Selling, distribution or promotional efforts were misdirected	15.9	23.9
The product was the same as competing products ... a 'me too' product	14.8	25.0
Did not understand customer requirements; product did not meet his needs or specifications	13.6	26.1
Selling, distribution or promotional efforts were inadequate	9.1	31.8
A similar competitive product was introduced	10.2	22.7
Were unable to develop or produce product exactly as desired	11.4	19.3
Competitors lowered prices or took other defensive actions	12.5	13.6
Timing was too late	8.0	13.6
No market need existed for this type of product	5.7	18.2
Timing was premature	6.8	13.6
Government action/legislation hindered the sale of the product	2.3	3.4

There are many texts which deal in detail with market analysis and measurement and it is assumed that these are well known to readers. However, it is believed that even if one were to follow the advice contained in these books there would still be a significant probability of failure for one or other of two basic reasons.

First, failure is defined in terms of not achieving a target sale volume within some prescribed period of time, usually determined on the basis of the time necessary to earn a satisfactory rate of return on the capital employed. The difficulty in applying such a criterion is that the great majority of managers tend to use a straight linear extrapolation when projecting future sales, despite the fact that all the available evidence on the sales performance of successful new products points convincingly to some form of exponential growth. The theoretical expression of such a phenomenon is the well known product life cycle concept which postulates that all products pass through a life cycle characterised by slow initial development. This is followed in turn by a period of rapid growth and a period of maturity or stability, whereafter sales will begin to decline unless management take positive steps to extend or even rejuvenate the mature phase. Such a product life cycle is reproduced in Figure 1, together with a straightforward linear projection of expected sales. From this figure it is abundantly clear that in the initial phases the new product will consistently underperform against the target - indeed, the gap between the two will increase. Depending upon the time scale involved, it seems quite likely that many managements will withdraw a product from the market because of its apparently deteriorating performance (the gap between 'projected' and 'actual') without ever knowing whether it would 'take off', in which case actual sales would later greatly exceed those projected.

Of course, this is the major criticism levelled against the PLC concept - it can only tell you what the sales performance of a successful new product will look like - it cannot tell you if any given product experiencing difficulty in penetrating a market would be successful if you persevered with it.

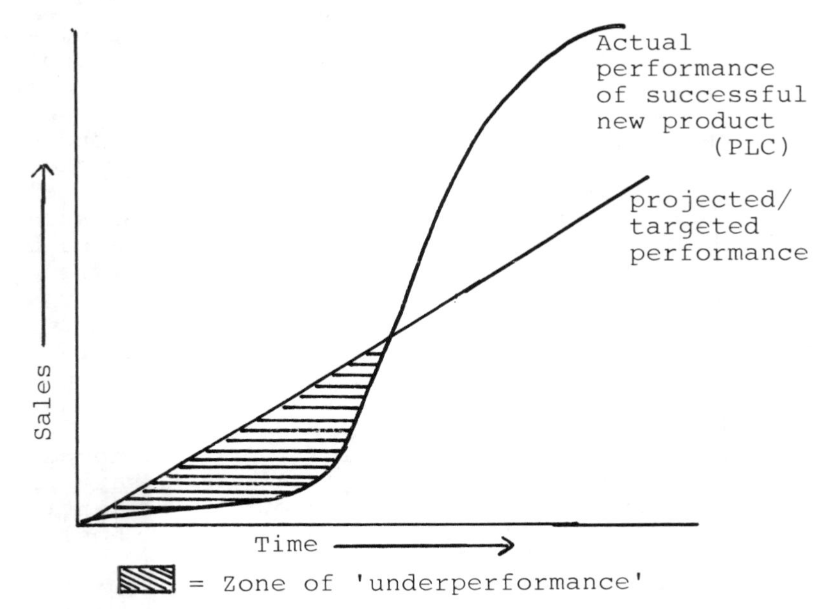

Figure 1. Product Life Cycle and Projection of Expected Sales

The second reason why existing advice on market analysis and new product development will never result in a 100 per cent record of success is that there is another important factor which has been omitted from conventional analysis and which accounts for a significant proportion of failures. This factor has been described in a number of ways, the most familiar of which is 'resistance to change'. But, however we describe it, it is an expression of customer attitudes towards new products, and it helps to explain why such products make slow initial progress when first introduced to a market.

It is vital that managers responsible for launching new products should understand why prospective customers may be slow to appreciate or accept the benefits claimed for such innovations. Clearly, if one possessed such information then it might be possible to devise strategies to overcome these difficulties. Basically, the problem seems to be that while most managers feel that they can pre-identify the most receptive market for an innovation, the evidence is that by and large they are not very good at it. With the benefit of hindsight it is not difficult to see why company A* could not seem to make up its mind about the benefits of adoption while company B accepted it almost immediately. The question is can we identify any patterns in the reactions of companies (or individual consumers) to new product propositions which would enable us to pick out the B companies in advance? The importance and value of attempting to do this is predicated by the consistency of 'S' shaped product life cycle or diffusion curves.

Elsewhere (Market Development, Penguin, 1983) I have examined extensively the evidence which supports the view that sales of most successful new products follow an S-shaped diffusion curve to the extent that one may infer the existence of a fundamental and logically necessary process. Of course, one can find exceptions - particularly in the case of fad or fashion goods whose sales take off almost vertically and decline just as quickly - but the regularity of the S-shape product life cycle curve indicates that in the vast majority of cases the sellers of new products will initially experience considerable difficulty in persuading prospective customers to try the new offering. Basically, this is because nearly all new products are substitutes for established products which are perceived as satisfactory by their users. While they are tried and trusted the new offering is an unknown quantity and so represents an unnecessary risk to the average user who would prefer to "wait and see" before taking a decision to buy the new product. On the other hand, some persons are more willing to take risks and may well derive considerable personal satisfaction from being the first to try the new thing and being seem as "innovative".

Hence, if one could identify in advance those potential users so predisposed one could concentrate one's marketing efforts on this most receptive sub-segment, thereby achieving initial sales more rapidly than would be the case if a

---

*While the ultimate success or failure of a new food product will depend on individual consumer preferences, it is the institutional buyers of the major supermarket chains who decide whether or not it will be offered for sale.

random approach were followed which presumes that all prospective users are equally likely to adopt first. The impact of only halving the time to first adoption is readily apparent from Figure 2.

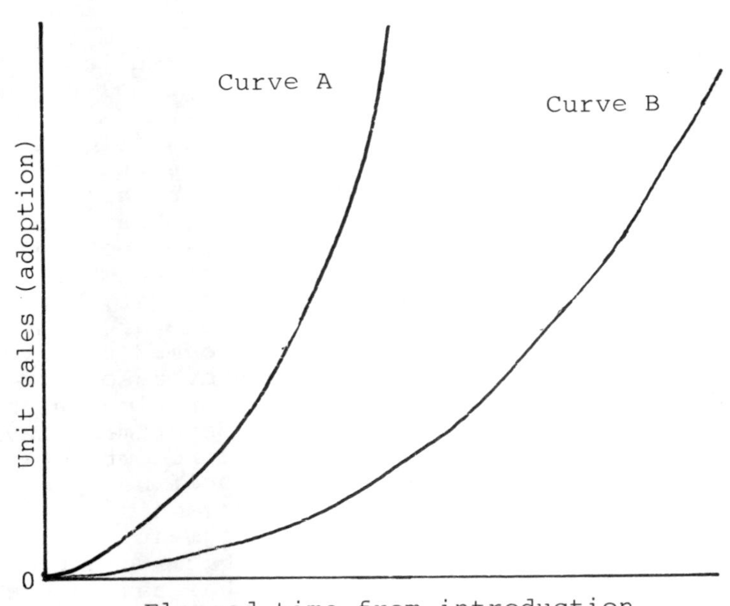

Curve A indicates exponential growth of sales if the first adoption is secured twice as quickly as the first sale in Case B.

Figure 2. Graphical Representation of Long-Run Impact of Halving Time to Secure First Adoption of an Innovation

In reality, of course, the assumption that all prospective users are equally likely to adopt is rarely applicable and most marketing plans are based upon some form of market sub-division or 'segmentation'. In essence, the concept of market segmentation rests upon recognition of a differentiated demand for a product while its use as a marketing tool depends upon identification of the most appropriate variable or variables with which to subdivide total demand into economically viable segments. (In this context 'economically viable segment' may be interpreted as 'being of sufficient size to enable a marketer to earn an adequate profit by catering to the specific needs of its members'). Thus, in practice, a major concern of the marketer when planning the introduction of a new product is determination of which segment has the strongest demand for the innovation, and what are the most suitable policies for converting this demand into actual purchases.

It is difficult to assess just how effective such market segmentation strategies are but, as the incidence of new product failure clearly indicates, they meet with mixed success. In part, some failures are attributable to neglect of generally accepted marketing principles for which a remedy is immediately available. For the rest, however, it is reasonable to assume that there are many situations where a product is deemed to have failed because its sales have not achieved some predesignated level after a given time which, in turn, is due to the marketing department having incorrectly identified the most receptive market segment. Therefore, reduction of new product failure is dependent upon improved identification of the most receptive market segment, and the most receptive individuals within that segment. Thus, while present segmentation techniques such as four-figure SIC analysis permit broad identification of the most likely segment, we need additional techniques to help us pinpoint specific prospects.

If the Product Life Cycle does reflect reality - and I am convinced it does - then one might inquire why it is that new products are slow to penetrate markets? The short answer is 'resistance to change'.

RESISTANCE TO CHANGE

Eric von Hippel of MIT offers a useful typology of innovation when he suggests three broad categories:

1. Known need;
2. Customer active or need pull;
3. Supplier active or technology push.

Observations and the limited evidence available (von Hippel only located eight studies of the customer active paradigm) suggest that this ordering reflects increasing resistance to change and, therefore, an increasing probability of failure due to delays in achieving an acceptable sales volume.

Known need innovations are instantly recognisable as 'just what I've always wanted'. ICI's synthetic pyrethroid insecticide Ambush fell exactly into this category in that it offered a whole cluster of desirable benefits - wide spectrum of activity, remarkable knock- down, a relatively fast good kill combined with very low mammalian toxicity and non-persistence in the environment. Undoubtedly, its very rapid market acceptance also owed much to careful preidentification of the best target markets and imaginative marketing, but, fundamentally, its success may be ascribed to the best advice of all to an innovator - build a better product at an equivalent price, or an equivalent product at a lower price.

Needled woven polypropylene exhibited exactly the same instantaneous recognition when it was first offered to tufted carpet manufacturers, as it offered the opportunity to reduce the pile weight of the carpet without fear of the backing material 'grinning' through. The economies offered were significant and nearly every manufacturer with brands designed for the low price end of the market adopted the new material immediately.

Need pull innovations also encounter relatively little resistance for, by definition, they represent a response to a known and declared interest. Frequently such innovations are the direct result of an approach by a user to a prospective supplier and are the outcome of joint development work. This is not to say that the innovator will not encounter any difficulties when they seek to offer the new product to a wider market, but it is usually much simpler to make adjustments to a product which exists and can be shown to work in analagous situations than is the case with a completely new and untried product. A recent example of this which came to my attention was the use of eddy current methods to detect cracks in fasteners used in aircraft construction. The innovator had developed a successful piece of equipment for the RAF under contract, but was surprised to find only limited interest from civil airlines when he tried to sell it to them. Analysis of this lack of interest made it clear that the main problem was that the innovator was seeking to 'push' his innovation on the civil airlines on the assumption that if it was good enough for the RAF it was good enough for them. In fact, it was too good and provided a level of performance far beyond that required, but at the sacrifice of speed which was seen as essential to keep planes in service. Once this fact became known it was a relatively simple modification to increase the speed of scan and desensitise the accuracy to an acceptable level.

Thus, it is with the third category - the supplier active or technology push innovation - that most difficulties are encountered for, in this case, prospective customers have envinced no open or explicit interest in a new product. Such an approach is usually characterised as 'production orientated' and conforms with the stereotype of the lone inventor single mindedly pursuing his goal oblivious to the world outside. While such an approach frequently does lead to the creation of products which are 'new' in the technical sense, the innovator then has to identify potential users whose needs match the benefits to be conferred from acquisition of the new product. There is much evidence to suggest that it is here that there is the greatest potential for failure due to a mismatch in the perception of supplier and user - the intervening variable to which reference was made earlier.

If prospective users are unaware of a need for an improved product to substitute for something in current consumption, then it seems reasonable to assume they are satisfied with the existing product. If we were to undertake a full appraisal of everything which might conceivably substitute for present behaviour, the information overload would become intolerable but, fortunately, we have a defense mechanism of selective perception which suppresses or screens out irrelevant or redundant information.* It follows that unless we can couch information about our new product in terms which are meaningful to prospective customers then they will never even be aware of it at the conscious level. Sophisticated marketers are well aware of the possibility and devote considerable effort to try and pre-identify the most receptive market

---

*A study by Bauer & Greyser in the USA indicated that we perceive only 1% of the advertising messages to which we are exposed. Of this 1% many will be intrusive and consciously suppressed.

segment, ie. the one with the least resistance to change. Surprisingly, they are often very poor at this and my own research suggests that this is because they give too much weight to the most attractive segment in terms of potential sales volume rather than the segment with the most pressing need. On reflection, however, and with the benefit of a little hindsight, it is clear that the customers with the biggest potential need are likely to have the greatest commitment to the existing technology and, therefore, the greatest inertia to overcome to make a change.

DEVELOPING A STRATEGY FOR NEW PRODUCT DEVELOPMENT

Thus far we have reviewed the reasons why new product development has become the dominant competitive strategy in most industries/markets; why many, and probably the majority of new product introductions are commercial failures; and we have suggested that the regularity of the diffusion curve for new products (or product life cycle) is such that it offers spectacular rewards to those who can pre-identify the innovators as they will accelerate the take up of the new product by legitimating it in the eyes of their more conservative cousins. The question remains how can we put these insights to use in the context of developing new food products?

First, it is important to recognise that in competitive markets one cannot afford to do nothing - just to retain one's existing customers it is necessary to pursue a policy of continuous product improvement. But, if one has ambitions to grow then one must innovate and here one has basically two options - expand the market and/or change the technology. These options are summarised in Igor Ansoff's Growth Vector Matrix (Figure 3) which shows that one can grow through product development or market (customer) development and that where one achieves both together then a diversification has occurred and you are in a completely new business.

While the matrix is helpful in spelling out the broad options of technology push versus market pull it suffers from the defect that many people interpret this as offering two mutually exclusive alternatives usually described as a production or marketing orientation. In truth, successful innovation will only occur where technology and marketing are properly integrated. In some

MARKET \ PRODUCT	CURRENT	NEW
CURRENT	MARKET PENETRATION	PRODUCT DEVELOPMENT
NEW	MARKET DEVELOPMENT	DIVERSIFI-CATION

Figure 3. Ansoff's Growth Vector Matrix.

instances the initiative will come from the laboratory, in others it will be derived from specific customer requests for a product modification or extension but, in all cases, a profitable outcome demands the combination of both. Such combination is implicit in the normative model of new product development which is usually represented as in Figure 4. Thus, Ideas can emanate from anywhere - the purpose of Screening is to establish which offer the strongest prima facie likelihood of success in terms of the market opportunity related to the company's competitive strengths. Those short-listed will then be subjected to detailed Business Analysis to determine which will be put into the physical development phase.

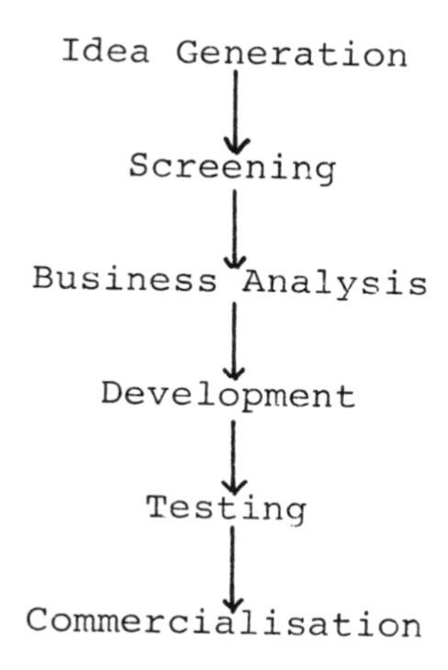

Figure 4. The Normative NPD Process.

In the case of food products Testing assumes a much greater importance than is the case with many industrial or consumer durable goods for the simple reason that ultimately "performance" will be measured on subjective sensory factors as opposed to objective outputs/outcomes appropriate to products which are not physically consumed by the user. This is not to say that one cannot usefully classify consumers in terms of subjective sensory preferences, only that it is particularly important for food products and that the number of failures would seem to suggest that the efforts of this particular branch of the Society of Chemical Industry are not as fully appreciated or used as they should be. It may also reflect a belief in scientific differences which are neither apparent nor important to final users. As an observer I suspect that the opportunity for improved performance (success) depends most on the screening and testing phases if for no other reason than that the costs involved begin to escalate exponentially once one becomes committed to production on a Commercial scale.

At this juncture it may be helpful to state the obvious. Consumers are motivated by their own self-interest, they don't owe you a living, they expect you to anticipate their needs and, because you have done so in the past, they are motivated by continuously increasing expectations. If you meet these expectations they will maintain their loyalty - if you don't they will transfer their patronage to someone else who can. It is this transfer of loyalty which leads to the decline of one product's life cycle and the initiation of another new one - after all there is a finite limit to how much we can physically consume and developments in food products invariably depend upon substitution for existing products. But, because of our intrinsic resistance to change, the existing suppliers have a natural advantage provided they practice a policy of continuous product improvement and show themselves to be responsive to changing consumer needs. Brand names become synonymous with strongly held expectations of performance, quality and satisfaction and provide a strong competitive advantage to the firm which pursues a strategy of continuous incremental innovation.

If, therefore, you wish to break into an established market there are, basically, only 2 options available to you. Either you introduce a radical innovation which makes existing solutions to the consumers needs obsolete (extremely rare, difficult and expensive) or you acquire an established participant with brand loyal customers (expensive but not too difficult qv. Nestle and Suchard's bid for Rowntree). As noted the problem with radical innovation is that not only is it difficult to accomplish technically but it usually encounters the strongest resistance to change due to its novelty and contrast with accepted practice. Accordingly, and with the exception of strongly desired innovation, such as a cure for the common cold, which everyone wants, most firms would be well advised to concentrate upon a policy of incremental innovation to protect and strengthen their existing markets. This us not to argue against basic research nor to ignore the potential of a breakthrough discovery if one is made;. rather it is the counsel of commercial common sense which recognises that a dependable earnings stream is necessary to underpin both basic research and radical innovation.

It is probably also true that marketing has little role to play in influencing the direction of basic research - few consumers are able to articulate the properties of radically new products they would like. The best you can hope for is that they'll recognise something good when they see it! It follows that if a radical innovation is achieved then marketing's contribution will be to match customers to the product - the obverse of the usual application of marketing which requires that one develop products to meet specified consumer needs. In both cases, however, the major contribution of marketing to new product development is likely to arise from its ability to identify and classify consumer needs for it is a lack of understanding of these which underlies most of the causes of failure reviewed earlier.

The techniques and approaches available for identifying consumer needs and using these as the basis for developing differentiated marketing strategies are numerous and complex and far beyond the scope of a paper of this kind. However, two examples of the application of such techniques as an aid to identifying marketing opportunities/ defining markets may help establish the contribution which marketing research has to offer technological research.

PROBLEM INVENTORY ANALYSIS AND BENEFIT SEGMENTATION

The two techniques selected to demonstrate the contribution of marketing research to NPD enjoy self explanatory titles and are easy to comprehend. Problem Inventory Analysis is a method for generating new product ideas by asking consumers to relate possible problems to specific products in which the researcher is interested, An example of the application of the technique to the food industry is provided by Figure 5 in which problem areas are grouped under five headings. By inviting consumers to identify which products have what problems one is able to discover opinions of both one's own and one's competitors products as well as elicit desired features not satisfied by current offerings.

Benefit segmentation is a technique developed by Russell Haley which he described as "... an approach based upon being able to measure consumer value systems in detail, together with what the consumer thinks about various brands in the product category of interest. While this concept seems simple enough, operationally it is very complex. There is no simple straightforward way of handling the volumes of data that have to be generated. Computers and sophisticated multivariate attitude measurement techniques are a necessity." The application of such methods to an analysis of the U.S. toothpaste market yields a readily understandable classification into 4 major segments with descriptive characteristics as shown in Figure 6.

SUMMARY

To sum up, it is clear that new product development is an increasingly necessary and increasingly risky business. While technological research will undoubtedly continue to yield a flow of potential new products, commercial success will still depend upon positioning them carefully so that their marketing is focused upon those consumers who will receive the maximum benefit from their consumption. Given the need to specify precisely what it is that different groups of consumers (market segments) want of a product it is common sense to advocate that technical development is more likely to be successful if one first determines customer wants and then develops the products to meet these wants. Either way, whether one pursues a push or a pull

Physiological	Sensory	Activities	Buying Usage	Psychological/Social
A. Weight - fattening - empty calories	A. Taste - bitter - bland - salty	A. Meal planning - forget - get tired of it	A. Portability - eat away from home - take lunch	A. Serve to company - would not serve to guests - too much last minute preparation
B. Hunger - filling - still hungry after eating	B. Appearance - color - unappetizing - shape	B. Storage - run out - package would not fit	B. Portions - not enough in package - creates leftovers	B. Eating Alone - too much effort to cook for onself - depressing when prepared for just one
C. Thirst - does not quench - makes one thirsty	B. Consistency/texture - tough - dry - greasy	C. Preparation - too much trouble - too many pots and pans - never turns out	C. Availability - out of season - not in supermarket	C. Self-image - made by a lazy cook - not served by a good mother
D. Health - indigestion - bad for teeth - keeps one awake - acidity		D. Cooking - burns - sticks	D. Spoilage - gets moldy - goes sour	
		E. Cleaning - makes a mess in oven - smells in refrigerator	E. Cost - expensive - takes expensive ingredients	

Source: Edward M. Tauber, "Discovering New Product Opportunities with Problem Inventory Analysis," Journal of Marketing, (January 1975), *69*, Reprinted from Journal of Marketing, published by the American Marketing Association.

Figure 5. Problem inventory analysis.

Segment name:	The sensory segment	The sociables	The worriers	The independent segment
Principal benefit sought:	Flavor, product appearance	Brightness of teeth	Decay prevention	Price
Demographic strengths:	Children	Teens, young people	Large families	Men
Special behavioral characteristics:	Users of spearmint flavored toothpaste	Smokers	Heavy users	Heavy users
Brands disproportionately favored:	Colgate, Stripe	Macleans, Plus White Ultra Brite	Crest	Brands on sale
Personality characteristics:	High Self-involvement	High sociability	High hypochondriasis	High autonomy
Life-style characteristics:	Hedonistic	Active	Conservative	Value-oriented

Source: Russell I. Haley, 'Benefit Segmentation ...', Journal of Marketing, vol. 32 (July 1968).

Figure 6. Toothpaste market segment description

strategy, success will accrue more often to the firm which integrates elements of both into the new product development process.

23

Consumer optimization of blended whiskies using the Simplex approach

A. A. Williams
Sensory Research Laboratories Ltd., 4 High Street,
Nailsea, Bristol, BS19 1BW, UK

ABSTRACT

Manufacturing whiskies for the mass market inevitably involves the blending of single spirits, in order to achieve products with the desired flavour characteristics.

The Simplex algorithm coupled with the power of the modern micro computer provides a means of rapidly determining the amounts of ingredients within a product which best satisfy the consumer, hence providing a beverage producer with valuable information when finalising the composition of his product.

The paper outlines the approach and illustrates with an example, how it might be applied to whisky blending.

INTRODUCTION

The final composition of any blended whisky is decided by the blender using all his or her knowledge and experience, but what checks are there that the views of such people truly reflect current trends in the market place.

The Simplex algorithm offers the potential of rapidly homing in on individual consumers' ideal blends and of providing some indication as to the effect, on acceptance, of changing the composition of the product around this ideal. Such information is invaluable to a blender when finalising his or her product.

OPTIMIZATION USING THE SIMPLEX APPROACH

Recent developments which provide the ability for assessors to interact with a computer in a controlled environment (Williams and Brain 1986:), has opened up new approaches in the field of sensory analysis. The speed of operation of the microprocessor enables calculations to be performed as an assessor is evaluating samples so that subsequent to the first of samples, the most appropriate products are always presented to the assessor to achieve the objective of the test.

In the Simplex approach to product optimization a matrix of samples covering the variables to be optimized needs to be available to the assessor (these can either be premixed or blended as required). The assessor is presented with one of the samples to which he gives a score reflecting its acceptance (low = poor, high = good). He or she is then presented with a second sample providing a contrast with the first, which is again scored in respect to acceptance, the number given reflecting its acceptance relative to the first sample assessed. After the assessor has been presented with one sample more than the number of variables being investigated, the computer program selects the next most appropriate sample to be presented in order to rapidly home in on the assessors ideal product.

THE SIMPLEX ALGORITHM

The Simplex method (Nedler and Mead 1965: O'Neill 1971) provides a means of moving through a matrix according to a set of defined rules. The Simplex itself is a simple geometric figure, one dimension more than the number of independent variables being optimized. In the example given (figures 1-3) in which the final composition of a three whisky blend is being optimized this is a triangle (the proportion of the third whisky being defined by the amounts of the other two). In its simplest form the Simplex algorithm uses the scores given to the first three samples and constructs a new simplex by projecting the least liked sample through the line joining the the most liked pair of samples, to identify the composition of the next sample to be presented (figure 1). A new Simplex is then

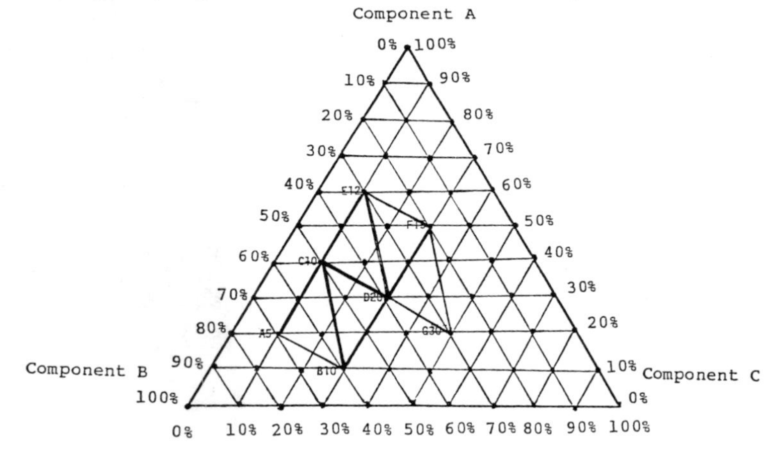

Figure 1 The movement of a simplex through a sample matrix

constructed based on the scores given to the last three samples assessed and the exercise repeated until the optimum is reached. Rules provide contingencies for samples being projected outside the matrix and to prevent a sample projecting back onto itself.

EXAMPLE

Typical results from an individual assessor evaluating a theoretical blend of three whiskies (a base Speyside malt, a Highland malt or malt blend with distinctive flavour characteristics and a blend of grain whiskies) is given in figure 2. In this example, although the full range of blends which could have been presented was 66 a particular individual usually targets in on his or her ideal formulation within six to eight assessments.

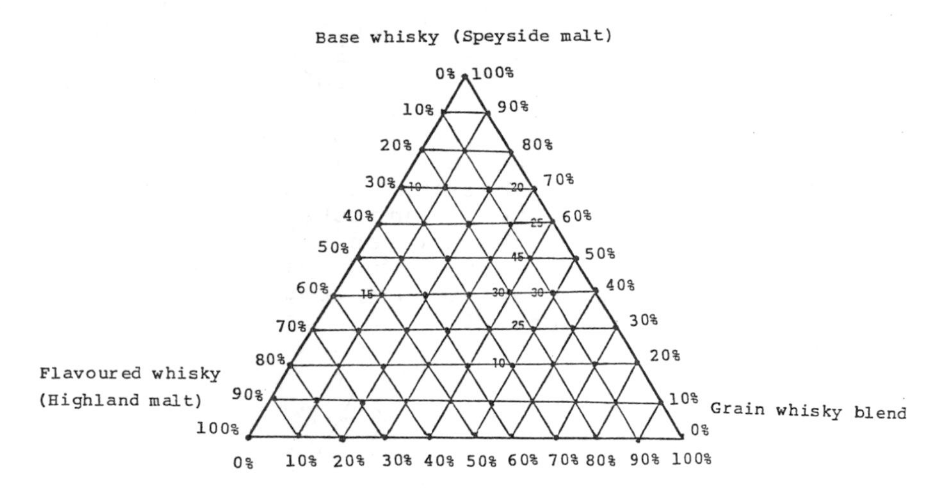

Figure 2 Assessment scores given by a single assessor

Information from individual assessors may be combined in various ways to provide a picture of the response of the population as a whole. Figure 3 is

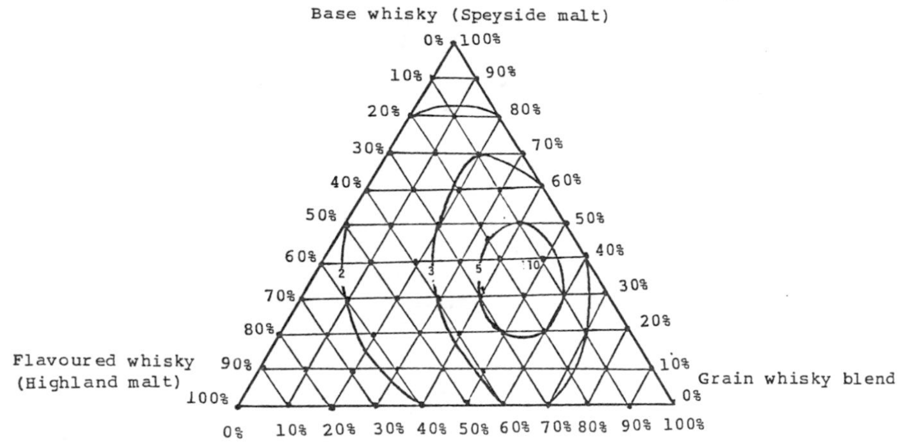

Figure 3 Contour map based on the number of assessors giving ideal products at a given composition

a contour plot based on the number of assessors having their ideal product at a particular blend composition.

REFERENCES

Nedler J.A. and Mead R. A Simplex Method for Function Minimisation. Computer Journal 1965, 7 308-13.

O'Neil, R. Function Minimisation using a Simplex Procedure. Applied Statistics 1971, 20, AS 47.

Williams, A.A. and Brain, P. The Scope of the Microcomputer in Sensory Analysis. Chem and Ind 1986, 118-122.

24

Legislative aspects of flavour use

M. E. Clowes
Regional Food Technology Centre, PFW (UK) Ltd.
PO Box 18, Greenford, Middlesex, UB6 7JH, UK

INTRODUCTION

In this paper I intend to deal with the position regarding flavourings with respect to the European Common Market (1,2) the recent MAFF News Release on the Consultation on the Use of "Natural" in Food Labelling (3), and other National regulations, with reference to the distilled beverage industry.

Before dealing with these subjects it is necessary to clarify some basic facts about what flavourings are. It is not adequate to say that flavourings are preparations which are added to food to give it flavour, since most compounded foods made up solely of food ingredients develop flavour as a result of processing. Also the use of herbs and spices, onion or garlic as part of a food recipe is primarily to provide flavour to the finished food. In no way do either of these circumstances indicate that flavourings have been used. My prefered definition is that developed by the International Organisation of the Flavour Industry (IOFI) -Flavourings are "Concentrated preparations, with or without flavour adjuncts, used to impart flavour, with the exception of only salty, sweet or acid tastes. They are not intended to be consumed as such" (4).Indeed if they were, they would be very unpleasant as they are around 1000 times more concentrated than normal foods. Flavour adjuncts include solvents and supports, antioxidants, sequestrants, preservatives, emulsifiers and stabilisers and so on.

Having decided what is meant by flavourings it is now possible to look at their constitution in more detail.

There are fundamentally two kinds of flavourings:-

(a) Defined flavourings: where the final preparation is a mixture of indivdual flavouring ingredients, of more or less the same chemical species as were added. That is they have not undergone any chemical change.

(b) Non defined flavourings: where the final preparation is obtained by deliberately reacting together the ingredients to provide a complex mixture of chemical substances of different nature to the original constituents.

As a food analogy, the addition of herbs and spices to food would correspond to a defined flavouring, whereas the flavour developed on cooking meat would be non-defined. There are of course flavourings that are a mixture of these two basic types, in the same way as in a food such as beef curry.

Defined flavourings

Defined flavourings are most easily dealt with by reference to the individual flavouring ingredients which are mixed together to produce the finished blend. Often such blends are dissolved in an appropriate food solvent, made into an oil in water emulsion or added to a solid carrier in order to produce a preparation of a suitable concentration to be capable of being efficiently used by the food manufacturer. In terms of flavouring legislation these solvents or carriers are not considered as part of the flavouring although, of course, they have to be permitted in the intended country of use.

The flavour industry internationally recognises three types of defined flavouring ingredients:-

The first of these is:

(a) Natural - That is flavouring ingredients prepared from animal or vegetable raw materials by the use of only physical and/or microbiological methods.

In this category there are around 400-500 flavour materials. Commonly occuring sources of natural flavour ingredients are extracts either aqueous or solvent extracts of fruit juices, coffee, vanilla, ginger and so on. Essential oils obtained by steam distillation or extraction are good natural sources, whilst distillates, tinctures, absolutes and isolates can be obtained from many common plants and herbs.

The second type is:

(b) Nature Identical -These are defined substances chemically identical to substances which occur in materials commonly used as food.

In this category there are over 3000 flavouring materials.

The third type is:

(c) Artificial -This is defined flavouring substances not yet identified in a natural product intended for human consumption.

In this category there are around 200 flavour materials. This number is getting smaller as research identifies more of these materials in foods.

To put the matter of flavour use in perspective, Figure I shows the actual amounts of flavour used in the UK. It shows that 86% of flavour ingredients sold are in fact natural food substances, solvents and supports,9% are process flavourings and only 5% are flavouring substances.

INGREDIENTS OF FLAVOURINGS AS SOLD BY WEIGHT (TOTAL U.K. MARKET – 50,000 TONNES/YR)

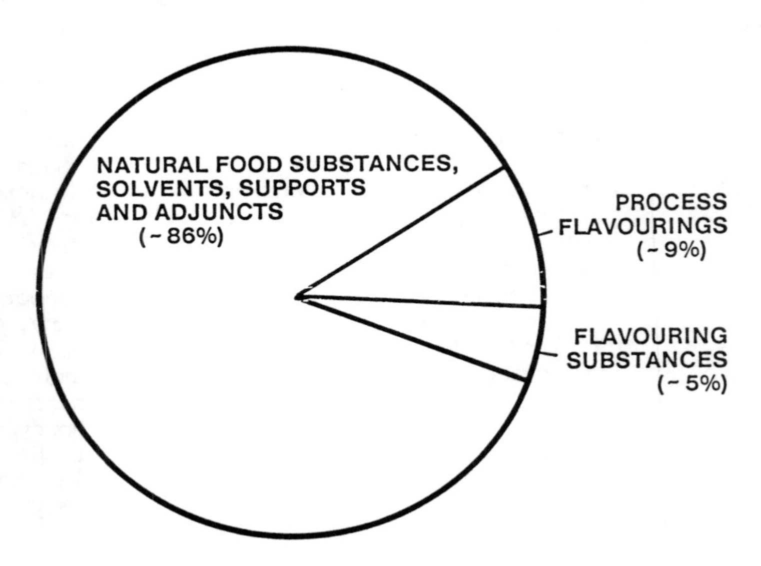

Figure II shows the proportion by weight of the 5% flavouring substances used. It can be seen that of these 90% are natural materials, 9% are nature identical and only about 0.7% are artificial flavouring substances.

FIGURE II

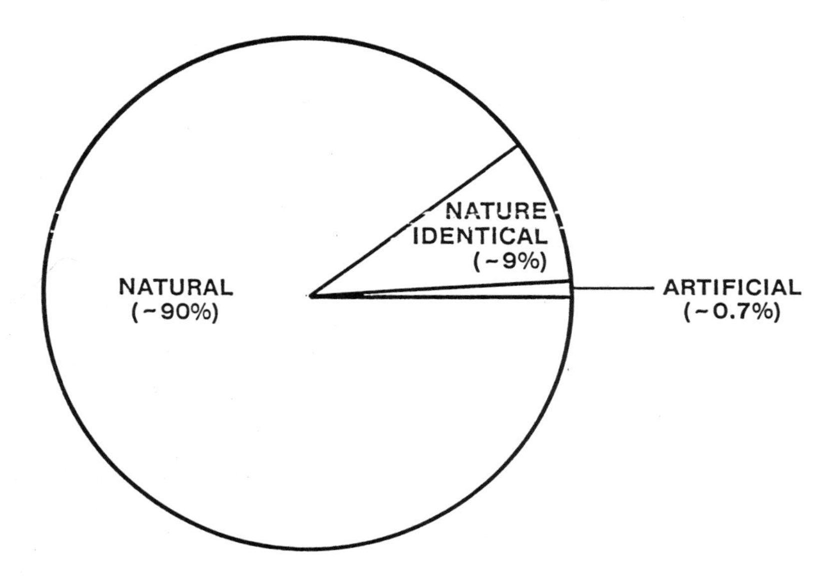

Flavourings containing more than one of these categories are defined by the lowest category which is present. Thus a flavouring containing mainly natural ingredients but with a least one nature identical ingredient would be termed "nature identical flavouring" or if at least one artificial is present, the whole flavouring becomes artificial. Solvents, carriers or other non-flavouring ingredients are disregarded in the categorisation of finished flavourings. This method of categorisation has been adopted by a number of national legislations notably W.Germany, Italy, Holland, Finland and Spain. However, USA and Japan recognise only two categories in their legislation, natural and artificial, the latter representing an amalgamation of the nature identical and artificial categories. However, the USA definition of natural is somewhat wider than usually accepted in Europe, so that some materials which are regarded as nature identical in Europe maybe regarded as natural in USA.

Non-defined flavourings

It is difficult to provide an exhaustive list of this type of flavouring, since many of these are at the cutting edge of development in the flavour industry. However, some of the better known types are as follows:-

(a) Process flavourings: the most important of these are those that are analogous to the reactions which occur in cooking meat. The essential ingredients are proteins or amino acids, a reducing sugar and a sulphur donor, usually cysteine. The mixture is heated under defined conditions to produce a flavour reminiscent of meat. Similar flavourings reminiscent of chocolate, malt and coffee have been prepared or proposed. Ultimately it might be possible to prepare a single precursor for the characteristic constituents of these flavourings, which, when it is processed under specified conditions, will produce the required flavour in the finished food.

(b) Enzymatic and/or Microbiological flavourings are produced by the modification of a food, or food derived substrate by fermentation. This has been a technique used for producing dairy flavourings, and more recently for the preparation of natural specified substances such as butyric acid or acetaldehyde.

(c) Pyrolysis flavourings: By the controlled dry or steam distillation of hard woods, smoke type extracts can be prepared, which are analogous to the traditional smoking of foods, and are used for this purpose. They have the advantage over the traditional method in that they can be treated to reduce the polycyclic carcinogens present in natural smoke to insignificant levels.

Having looked at the background to flavourings, it is possible to look at the European Draft Directive and its present position. The original Commission proposal (5) was submitted to the European Parliament in 1980 and modified in 1982 (6,7) The European Council achieved a common position on this framework Directive on 22nd June 1987 and it was submitted to the European Parliament (1). It passed its second reading on 9th March 1988 when 13 modifications were proposed (2), 7 of which have been accepted by the Commission (8). Most of these are of an organisational nature, but one, amendment (11), is of significant technical and commercial importance. I will return to this in the review of the document. I now see no reason why the Directive should not pass its third reading without further significant modification and become Community law.

THE EUROPEAN FLAVOUR DIRECTIVE

Whilst it is not possible in the time available to go through each clause in detail, I intend to draw attention to those which are likely to have a significant effect on the flavour and food industries. Firstly Article 1 which deals with definitions which are somewhat similar to those proposed by IOFI, except for the "flavouring" definition. Art 1.2 (a) "flavouring" means flavouring substances , flavouring preparations, process flavourings, smoke flavourings or mixtures thereof.

Para (b) then proceeds to define natural, nature identical and artifical flavouring substances, but the proposal by the European Parliament to call them by these familiar designations was not accepted (8)

Para (c) defines "flavouring preparation" as a product, other than those in (b), with flavouring properties obtained by physical, enzymatic or microbiological processes from vegetable or animal materials either raw or processed for human consumption by traditional food preparation processes,

Para (d) defines "process flavouring" as a product of heating together a mixture of ingredients, not necessarily themselves having flavouring properties, of which at least one contains amino nitrogen and another is a reducing sugar to a temperature not exceeding 180°C and a period not exceeding 15 minutes.

This can be a limitation on developers as not all processes will neatly fit into this definition. For instance barbequeing a piece of meat reaches far higher temperatures than this, and many natural cooking processes exceed 15 minutes, so the definition is somewhat unrealistic.

Article 4 Limits the level in foodstuffs of a number of pharmacologically active constituents of natural flavouring preparations as shown in the Table 1.

TABLE 1
Maximum Limits in Food

Substance	ppm	Exceptions
Agaric Acid	20	100ppm in foods with mushrooms
Aloin	0.1	50ppm in alc. beverages
beta-Asarone	0.1	1ppm in alc. bevs & snacks
Berberine	0.1	10ppm in alc. beverages
Coumarin	2	10ppm in confections. 50ppm in Chewing Gum 10ppm in alc.beverages
Hydrocyanic Acid	1	50ppm in nougat & marzipan. 1ppm in alc. beverages 50ppm in canned stone fruit.
Hypericine	0.1	10ppm in alc. beverages 1ppm in confections
Pulegone	25	100 in beverage 250 in mint bevs 350 in mint confections
Quassine	5	10 in confections 50 in alc. beverages
Safrole & iso safrole	1	2 in alc beverages<25% 5 in alc beverages>25% 15 in foods with mace & nutmeg
Santonin	0.1	1 in alc.bevs>25%
Thujones	0.5	5 in alc bevs<25% 10 in alc bevs>25% 25 in fods with sage 35 in bitters

Examples of foods containing some of these substances are Fly Agaric which is a distinctive red and white mushroom which can make you very ill if consumed, hence Agaric acid is on the list.

Another source is Pennyroyal, related to mint, which contains Pulegone, much used in the past by ladies who found themselves in an unexpected state, as an abortifaciant. Thujone is found in the Sage plant.
I think looking at these substances it is fairly obvious why they were included.

Article 5 is probably the most important part of the whole Directive since it deals with the provisions which are regarded as appropriate to control each category of ingredient of finished flavourings.

Para (a) Lists 7 classes of sources and ingredients as follows:-

- (i) flavouring sources composed of foodstuffs, herbs and spices.
- (ii) flavouring sources composed of animal or vegetable non-food substances
- (iii) natural flavouring substances
- (iv) substances nature identical to foodstuffs, herbs or spices.
- (v) Substances nature identical to non-foods
- (vi) Artificial flavouring substances
- (vii) Source materials for the production of smoke or process flavourings and the reaction conditions for their production.

It does not specify how each category will be controlled although most legislators would prefer positive lists for each group. This is not felt to be appropriate by the flavour industry, since it would mean that little further research would be undertaken into the constitution of the flavour of natural foodstuffs. The results of this research would be available to one's competitors due to it having to be placed on a positive list. If positive listing were to become mandatory some method of confidential safety evaluation, maybe by National Authorities must be accepted and a period, of say five years, be available to the discoverer before publication, in order to allow him to capitalise on his research investment. Without such a period of grace natural product flavour research will collapse with the consequent loss to the consumer.

Paras (b) and (c) deals with flavour adjuncts such as solvents and supports and any special provisions for the protection of public health or trade.

Article 9
Para (2) is concerned with the designation of "natural" with respect to flavourings and has undergone substantial modification by the European Parliament (2,8). The present text allows that the word "natural" may only be used for flavourings in which the flavouring component contains exclusively flavouring preparations and/or natural flavouring substances.

Para (b) If the sales description of the flavouring contains a reference to a foodstuff or a flavouring source, the word "natural" may only be used if the flavouring component is wholly or mainly derived from the foodstuff or flavouring source concerned. You should note that the word "mainly" is one of the significant modifications. However, it still does not clarify what the position would be with for instance, "natural barbeque flavour" made entirely from natural ingredients. I will return to this in connection with the MAFF proposal on "natural".
The remainder of the draft directive is standard including the time scale for implementation.

It was intended that measures to comply with the Directive be adopted within 18 months, and prohibition of non-complying flavourings within 3 years. The Commission has decided that this is an impossibly short period to collect, evaluate, and make appropriate decisions on at least 3500 flavouring substances, and source materials. Maybe a more realistic period might be between 5 and 10 years, but this has not yet been specified.

MAFF PROPOSAL

Turning now to the MAFF consultation document on the use of "Natural" in Food Labelling (3): The recommendations contained in this document are derived from a survey commissioned by MAFF and carried out by the Local Authorities Co-ordinating Body on Trading Standards (LACOTS) (9).

The only comment on the criteria for flavourings is contained in para (c) on page 3 which states "Additives should not be described as "natural" unless they were produced from the relevant natural sources." This has been incorporated into MAFF proposals in para 6 (1) (c) as:-

"The term "natural" should be used without qualification only to describe flavouring substances obtained from recognised food sources by appropriate physical processing or traditional food preparation processes. Flavour(ing)s should only be so described when they are derived wholly from the named food source."
It is not clear what "without qualification" means in this

context. Does it mean that a raspberry flavouring made solely from natural flavouring materials, not entirely from raspberry may be designated "Raspberry flavour natural flavouring" or maybe "natural flavouring - raspberry type" and again how can a product such as "Natural barbeque flavouring" be dealt with? Further, if this definition is compared with that contained in the latest modification of the EEC draft Directive it will be observed that the former is more restrictive. I suggest that if the EEC Directive becomes UK law it will be necessary to modify the final sentence of the proposal to read "Flavourings should only be so described when they are derived wholly or mainly from the named food source".

In some countries the present legislators have gone for specific positive or negative legislation. Those with a positive list are the USA and Japan ie: no flavouring ingredient is allowed unless it is on the list. Some countries operate a mixed list system with positive and negative lists, these are: West Germany, Italy, Holland, Spain, Finland and Columbia.
In the USA for example, flavouring materials have to be GRAS (generally recognised as safe) and FDA approved. To obtain FDA approval and GRAS status it is necessary to declare the ingredient, which tends to lead to lack of development in new flavour materials. No company is going to invest thousands of pounds of development if they have to declare their work for all to see and copy, at no cost to themselves.

What of distilled beverages?

In Europe at present for the use of flavours in liqueurs the situation is in Table 2

Table 2

Holland	Natural, Nature-Identical and Artificial flavours allowed, no declaration required.
Belgium	Natural, Nature-Identical and Artificial flavours allowed, no declaration required.
France	Natural and for Apricot, Pineapple and Banana "arôme reinforcée 1 o/oo à base de concentre 4 fois", declaration is required.
Germany	Natural and Nature-Identical flavours allowed, no declaration for Natural. For Nature-Identical and Ethyl Vanillin declaration required.

U.K.	All flavour ingredients are allowed, no declaration required
Norway	All flavour ingredients are allowed, flavour declaration on a Product Information Sheet.
Finland	Natural and Nature-Identical flavour ingredients and Artificials, listed in (I.O.F.I) the Code of Practice, are allowed. No declaration requirements.
Sweden	All Natural and Nature-Identical flavour ingredients are allowed. No declaration required.
Denmark	All flavour ingredients are allowed. declaration "Aromastoffer".
Italy	"Aromi Naturali" (Natural and Nature Identical) are allowed, declaration required.
Spain	Natural and Nature-Identical flavour ingredients and positively listed Artificials are allowed. No declaration requirements.
Portugal	All flavour ingredients are allowed. No declaration.
Switzerland	Liqueurs are mixtures of specified alcohol, water, sugars and properties-defining ingredients and flavourings. Berry-named liqueurs have to be produced with fruit or fruit-juices. Other liqueurs with fruit names like Bergamot, Banana and Pineapple, a little reinforcement of Artificial flavours is allowed. Cherry Brandy Kirschwasser and Cherry Syrup. require Natural Flavours.

The European court has ruled that a liqueur manufactured and offered for sale in one member country of the EEC must be free for sale in all member countries, providing it is not considered harmful by the importing country. It should be labelled in conformity with the law of the country.

In the USA the use of flavourings in spirits is governed by the BATF regulations. (The Bureau of Alcohol, Tobacco and Firearms). This states that to declare the flavour as Natural it is permitted to use up to 0.1% artificials. There are specific maximum levels allowed for Maltol, Ethyl Maltol, Vanillin and Ethyl Vanillin, which have to be declared to the BATF together with Natural or non Natural colours, and the amount of say Propylene Glycol used as the solvent. All materials used must be permissable according to FDA regulations.

In the USA the most dynamic growth in flavoured alcoholic beverages has been that of peach flavour. The start of the trend was DeKuyper Peachtree Schnapps, which went from 3% of the schnapps market share to over 20% in its first 18 months, and is now the major flavour taking over from peppermint Schnapps. This sparked off a wave of Peach flavoured drinks from Schnapps to colas, exotic cordials, wine and frozen cocktails, even peach flavoured vodka is now on sale. It has been estimated that some 8 - 10 million dollars of peach flavour is being sold to the US alcohol industry. This trend can be seen in Europe as well.
A trend to lower alcohol products in the USA and Europe leads to products with less alcohol bite, so as well as flavours like peach, other flavours such as alcohol enhancers and other flavour boosters may be needed.
Products in Europe designed as low alcohol liqueurs or mixer drinks can be seen everywhere. The traditional and quaility liqueur producer Monin have there Depigny range of low Follies" liqueur aimed at the 25 to 35 year age group.Cusenier have launched "Peach Boy" a strong tasting low alcohol product for cocktails and mixer drinks.Unicognac have launched their range of "Seduction Liqueurs" with low alcohol,in three fruit flavours, aimed at women who do not drink cognac because it is to strong.

Using the facts outlined in this paper it can be seen that flavours whether natural, nature identical or artificial, can be used in the traditional and new lower alcohol products in many European countries, often without being declared.

What of the future for producing and using flavourings? There has already been a start on the production of flavours or flavour materials by use of biotechnology. In fact flavours have been produced by fermentation for thousands of years in the form of beers, wines and spirits. What is needed is the production of those desirable flavours without the production of alcohol, possibly by choosing the correct substrate, or perhaps by gene manipulation. It could be that flavours might be produced from cell cultures, either as total flavour or individual flavour ingredients.
How about artificial ageing of alcoholic products? With the lack of numbers of sherry casks now for the ageing of whiskies, the full traditional flavour is not developed. Blends of whiskies from sherry and non-sherry oak casks have to be produced at present. It is possible using certain materials to impart an "aged" flavour to whisky and sherry and could possibly cut down on the ageing time needed.

These are just some of the possibilities for the future use and production of flavourings.

I hope in this paper I have been able to bring you up to date with respect to the state of play on flavourings legislation and give some pointers to the future of flavour production.

REFERENCES

1. Council Directive on approximation of the laws of the Member States relating to flavourings for use in foodstuffs and to source materials for their production. EEC document 7259/87 (22/6/87)

2. European Parliament decision on the Councils' common position for Flavourings Directive. European Parliament document J269A (8/3/88) PE 120.963pp 26-33

3. MAFF News Release. Consultation on the use of "Natural" in Food Labelling 386/87 (16/12/87)

4. Code of Practice for the Flavour Industry - International Organisation of the Flavour Industry - 2nd Edition 1985

5. Commission proposal for a Directive on flavourings. Official Journal No C144, 13/6/1980 p9

6 Opinion of the European Parliament Official Journal No C66, 15/3/1982 p117

7. Amended Commission Proposal. Official Journal No C103, 24/4/82 p7

8. European Parliament - verbatim report of proceedings - Strasburg 8/3/88 pp.40-41 (document J269B)

9. The use of the Word "Natural" and its Derivatives in the Labelling, Advertising and Presentation of Food. Report of a survey by LACOTS,HMSO 1987

25

Factors influencing yeast performance during ethanol production

G. G. Stewart
Labatt Brewing Company Ltd.,
London, Ontario, N6A 4M3, Canada

INTRODUCTION

The production of ethanol by microorganisms as a result of the fermentation of substrates such as sugars and starch is a process that predates history. The uses of ethanol can be divided into a number of categories: (1) potable ethanol in beer, wine, saké, cider and perry, a variety of fermented fruit juices, and distilled beverages such as whiskey, gin, vodka, brandy, rum and liquors; (2) solvent ethanol in the laboratory, pharmaceutical preparations such as tonics and cough syrups, as a solvent for hop constituents, and in cosmetics; (3) as a cosurfactant in oil-water microemulsions; (4) as an antiseptic and sterilant; and (5) as a fuel in automobiles either on its own or more usually admixed with gasoline. Nevertheless, by far the largest volume of ethanol produced via fermentation is employed for potable purposes. Consequently, brewing, viticulture and enology and distilled beverages are biotechnological industries that make a significant contribution to the economy of most countries around the world.

The production of fermentation ethanol for fuel energy is still a vogue topic in the U.S.A.,

Brazil, Argentina, Thailand, Malawi, the EEC and a number of other countries. However, the topic is not new because fuel ethanol has been produced in Brazil since 1933 and both Germany and Japan used it as a fuel source during the 1939/45 war. Indeed, there was a monograph published in 1922 entitled "Power Alcohol" authored by Morien and Williams (Publ. Oxford Technical Publications) which described the significant developments that had occurred in this field to that date! Of course, it is the escalation in world crude prices that occurred in the mid-1970's that brought about a renewed interest in the fuel ethanol concept (Lyons, 1981). Although the world crude price has moderated in the past few years, there is still interest in fuel ethanol as a renewable liquid fuel particularly as a blend with gasoline to act as an octane enhancer to replace lead derivatives (Sage, 1987).

The advantages of alcohol fuels can be summarized as: (1) alcohols are readily available liquids, produced and utilized within existing technologies - purification of alcohols for fuel use only involves a relatively simple distillation step, unlike petroleum fuels which require complex refinement procedures; (2) alcohols can be produced from a number of renewable resources - production can be developed to best suit a particular region; (3) alcohol fuels burn cleaner than gasoline and, as discussed above, can be used as an octane enhancer as a blend in gasoline to replace lead derivatives; and (4) alcohols produced from waste and inedible agricultural products can be efficiently employed as substrates for controlled fermentation to produce food (eg., SCP), chemicals, drugs, enzymes, natural flavourings, etc.

The major carbohydrate sources for ethanol fermentations are: (1) sugars from beet, cane and sorghum; (2) inulin, a fructose polymer (fructan) found in such plants as the Jerusalem artichoke, chicory and dahlia; (3) starch from corn, wheat, barley and other cereals and cassava, potatoes and other root crops; (4) cellulosic plant materials (consisting of cellulose, hemicellulose and lignin) from wood and wastes; and (5) by-product carbohydrates from processing such as waste sulphite liquor, whey, food industry wastes, etc.

In the production of fermentation ethanol, the

microorganism being employed should possess a number of important characteristics: (1) rapid and relevant carbohydrate fermentation ability; (2) appropriate flocculation and sedimentation characteristics; (3) genetic stability; (4) osmotolerance (i.e., an ability to ferment concentrated carbohydrate solutions); (5) ethanol tolerance and the ability to produce elevated concentrations of ethanol; (6) high cell viability for repeated recycling; and (7) temperature tolerance.

There are a number of microorganisms (yeast and bacteria) that produce significant [greater than 1% (w/v)] quantities of ethanol (Tables 1 and 2). However at the present time, in excess of 96% of the fermentation ethanol produced globally involves the use of the yeast *Saccharomyces cerevisiae* and its related species.

Table 1. Some yeast that produce significant [>1% (w/v)] quantities of ethanol and the major carbohydrates utilized as substrates.

Saccharomyces spp:
- *Saccharomyces cerevisiae* and *Saccharomyces uvarum (carlsbergensis)*
  glucose, fructose, galactose, sucrose, maltose, maltotriose and xylulose
- *Saccharomyces diastaticus*
  glucose, maltose, dextrin and starch (glucoamylase)
- *Saccharomyces rouxii*
  glucose, fructose, maltose and sucrose (osmophilic)

Kluyveromyces fragilis
  glucose, galactose and lactose

Candida spp:
- *Candida pseudotropicalis*
  glucose, galactose and lactose
- *Candida tropicalis*
  glucose, xylose and xylulose
- *Candida wickerhamii*
  glucose, cellobiose and xylose
- *Candida shehatae*
  glucose, galactose, lactose and xylose

Pachysolent tannophilus
  glucose and xylose

Pichia stiptis
  glucose, xylose and cellobiose

Schwanniomyces spp:
- *Schwanniomyces alluvius*
  dextrin/starch (glucoamylase and α-amylase) and cellobiose
- *Schwanniomyces castellii*
  dextrin/starch (glucoamylase and α-amylase) and cellobiose

Saccharomycopsis (Endomycopsis) fibuligera
  glucose, cellobiose and dextrin/starch (glucoamylase)

Table 2. Some bacteria that produce significant [>1% (w/v)] quantities of ethanol and the major carbohydrates utilized as substrates.

Zymomonas mobilis
glucose, fructose and sucrose
Clostridium spp:
- *Clostridium thermocellum*
  glucose, cellobiose and cellulose (thermophilic)
- *Clostridium thermohydrosulfuricum*
  glucose, xylose, cellobiose, sucrose and starch (thermophilic)

Thermobacterioides acetoethylicus
glucose, sucrose and cellobiose (thermophilic)

YEAST STRAIN REQUIREMENTS

When considering the requirements for an acceptable yeast strain for ethanol production, one must first decide whether potable or industrial ethanol is being produced. There are some similarities and some differences between the two sets of requirements. In the production of potable (beverage) ethanol (be it beer, wine or spirits): (i) the yeast strain should be genetically stable and able to be used over repeated cycles with no genetic change in the culture; (ii) it must ferment the medium (wort, must or mash) in a reasonable period of time to product 4-12% (w/v) ethanol; (iii) the fermented medium should be palatable with no off-flavours (i.e., no significant degree of autolysis, phenolic off flavours or volatile organo-sulphur compounds); and (iv) at the end of the fermentation, the yeast should normally be easily removed from the medium (whiskey manufacture being an exception) either by flocculation or with a centrifuge. This cropped yeast should have a low dead cell population such that it can be re-inoculated into a subsequent fermentation with confidence. In the production of industrial ethanol (solvent and fuel): (i) the yeast strain should be genetically stable with reproducible performance over a number of fermentations; (ii) it must be able to ferment the medium in the shortest possible time at a high incubation temperature and produce ethanol near to the theoretical yield value; (iii) palatability of fermented product is not a consideration; and (iv) if the yeast is to be harvested for reuse at the end of the fermentation, it should be in a highly viable state - the ethanol tolerance of the strain is critical - if recycling of the yeast is not required, the yeast plus fermentation medium can be fed directly into the distillation still (Stewart *et al.*, 1983a).

The manufacture of beer is a biological process whereby agricultural products, such as barley and hops, are converted into beer by control of the biochemical reactions in malting, mashing and fermentation. The first two processes together produce a medium known as wort, to this yeast is added and the fermentation is allowed to proceed. During fermentation the yeast cells excrete major end-products such as ethanol and carbon dioxide into the medium, together with a host of minor metabolites, many of which contribute to the overall flavour of the final beer.

The brewing industry has traditionally been conservative when dealing with yeast performance optimization. Rather than dwelling exclusively on a single fermentation product yield (eg., ethanol), the brewer has quite rightly placed greater emphasis on the flavour characteristics of the final product. Brewery fermentations are slow, nonagitated and at below ambient temperatures. Quicker fermentations at higher temperatures or under agitated conditions tend to produce severe off-flavours (due to a plethora of compounds including fusel oils, esters and sulphur compounds) in the beer. However, as has already been discussed, in the production of fermentation industrial ethanol, the formation of a potable product is not an issue and, as a consequence, the fermentations are usually conducted at, or above, ambient temperatures and are invariably agitated.

In this laboratory, the research focus has been on brewer's yeast strains but interest has also encompassed the production of industrial ethanol via fermentation and the stress effects exerted upon yeast during fermentation (i.e., ethanol, temperature, osmotic pressure, etc.) (Stewart 1987a). The requirements of an acceptable brewer's yeast strain can be defined as (Stewart and Russell, 1986a): "*In order to achieve a beer of high quality, it is axiomatic that not only must the yeast culture be effective in removing the required nutrients from the growth medium (wort), able to tolerate the prevailing environmental conditions (eg., ethanol tolerance) and impart the desired flavour to the beer, but the microorganisms themselves must be effectively removed from the wort by flocculation, centrifugation and/or filtration after they have fulfilled their metabolic role".* In general terms, there are two types of beer - ale and lager. Ale is produced with strains of the yeast species *Saccharomyces cerevisiae* with fermentation temperatures between 18 and 22°C. Lager is produced with strains of *Saccharomyces uvarum (carlsbergensis)* which, unlike *Saccharomyces cerevisiae*, are able to metabolize the disaccharide melibiose due to the presence of the *MEL* genes which code for the production and secretion of melibiase (α-galactosidase). Lagers are produced at fermentation temperatures between 8-15°C. Indeed, strains of *Saccharomyces uvarum (carlsbergensis)* exhibit lower maximum growth temperatures (ca. 34°C) compared to that of strains of *Saccharomyces cerevisiae* (ca. 37°C) thus affording another method (in addition to the melibiose fermentation test) for distinguishing between the two species (Stewart *et al.*, 1985a).

Over the years, considerable effort has been devoted, in this laboratory, to a study of the genetics of brewer's yeast strains [and industrial yeast strains in general (Stewart, 1981; 1987a,b)]. The objectives of this study have been two-fold: (i) to learn more about the genetic make-up of industrial yeast strains and (ii) to improve the overall performance of such strains via genetic and environmental manipulation with particular emphasis being placed on broader substrate utilization capabilities, increased ethanol production and improved tolerance to environmental conditions such as temperature, high osmotic pressure, ethanol and salt.

Improvement in the substrate specificity of industrial yeast strains is a major objective of many zymologists (Stewart, 1981). Progress in achieving this objective has been impeded for a number of reasons which include the fact that industrial yeast strains (including brewing strains and strains employed for the production of distilling ethanol) are not easily amenable to genetic manipulation by classical techniques (Fowell, 1969). There are a number of methods that can be employed in genetic research and development of industrial yeast strains. These include *hybridization, mutation* and *selection, spheroplast* (or *protoplast*) *fusion* and *transformation* and, associated with it, *DNA recombination*. Industrial yeast strains are often polyploid or even aneuploid and, as a consequence, do not possess a mating type and have a low degree of sporulation and spore viability rendering genetic analysis of such strains difficult. However, modifications in incubation temperature and pre-sporulation carbon source have significantly increased sporogenic capabilities in industrial strains, thereby facilitating recovery of viable spores and thus providing an opportunity to manipulate such strains by classical (hybridization) techniques (Bilinski *et al.,* 1986, 1987a,b).

SUGAR AND CARBOHYDRATE UPTAKE BY YEASTS

As a group of microorganisms, yeasts are capable of utilizing a broad spectrum of carbohydrates and sugars. Nevertheless, none of the yeast species isolated to date from natural environments have been found capable of utilizing all of the readily-available sugars and carbohydrates. *Saccharomyces cerevisiae* has the ability to take up and ferment a wide range of sugars; eg., glucose, fructose, mannose, galactose, sucrose, maltose, maltotriose and raffinose. In addition, the closely related species *Saccharomyces diastaticus* and *Saccharomyces uvarum (carlsbergensis)* are able to utilize dextrins and melibiose, respectively. *Saccharomyces cerevisiae* and its related species metabolize sucrose rapidly and the initial step in its metabolism is hydrolysis of the disaccharide to glucose and fructose by the action of the enzyme invertase which is secreted through the periplasmic membrane into the medium. The hydrolysis products are then taken up by the yeast where they enter the common metabolic pathways of the cell. Sucrose is an important fermentation substrate in brewing and other fermentation-based industries employing yeast (eg., wine, potable spirit, fuel ethanol and baker's yeast production). Consequently, a basic understanding of the regulatory controls exerted on the uptake of its hydrolysis products (i.e., glucose and fructose), when present in a fermentation medium separately, together, or when added as sucrose, is important.

There are sugars that cannot be metabolized by *Saccharomyces spp.* but can be metabolized by other yeast genera; for example, xylose by *Candida shehatae, Pichia stipitis, Pachysolen tannophilus* and *Candida steatolytica,* cellobiose by *Schwanniomyces castellii* and *Candida wickerhamii,* lactose by *Candida curvata, Kluyveromyces fragilis* and *Kluyveromyces lactis,* starch by *Schwanniomyces castellii* and *Saccharomycopsis fibuligera* and inulin by *Kluyveromyces marxianus.*

It is rare that hydrolysis of either starch or cellulose leads to a medium consisting of a single sugar. As a result, fermentation of such a medium requires that the microorganisms in question are able to metabolize a number of sugars either together or sequentially. Further, the repressive effects of one sugar on the uptake of another have a profound influence on both the rate and extent of fermentation. The repressing effects of glucose during a mixed sugar fermentation can be seen in the fermentation of brewer's wort by *Saccharomyces cerevisiae* and *Saccharomyces uvarum (carlsbergensis)* (Stewart *et al.,* 1983). Wort contains the sugars sucrose, fructose, glucose, maltose and maltotriose together with dextrin material. In the normal situation, brewing yeast strains are capable of utilizing sucrose, glucose, fructose, maltose and maltotriose in this approximate sequence although some overlap does occur, leaving the dextrins unfermented (Figure 1). As previously discussed, the closely related species *Saccharomyces diastaticus* is able to utilize wort dextrins (and other forms of soluble starch) due to the production of an extracellular glucoamylase (Figure 2).

A major limiting factor in the fermentation of wort is the repressing influence of glucose upon maltose and maltotriose uptake. Only when approximately 50% of the wort glucose has been taken up by the yeast will the uptake of maltose commence. In other words, in most strains of *Saccharomyces cerevisiae* and related species, maltose utilization is subject to control by carbon catabolite repression. In a similar manner, the presence of glucose will repress the production of glucoamylase by *Saccharomyces diastaticus*, thereby inhibiting the hydrolysis of wort dextrins and starch. (Stewart, 1981; Stewart *et al.*, 1985b).

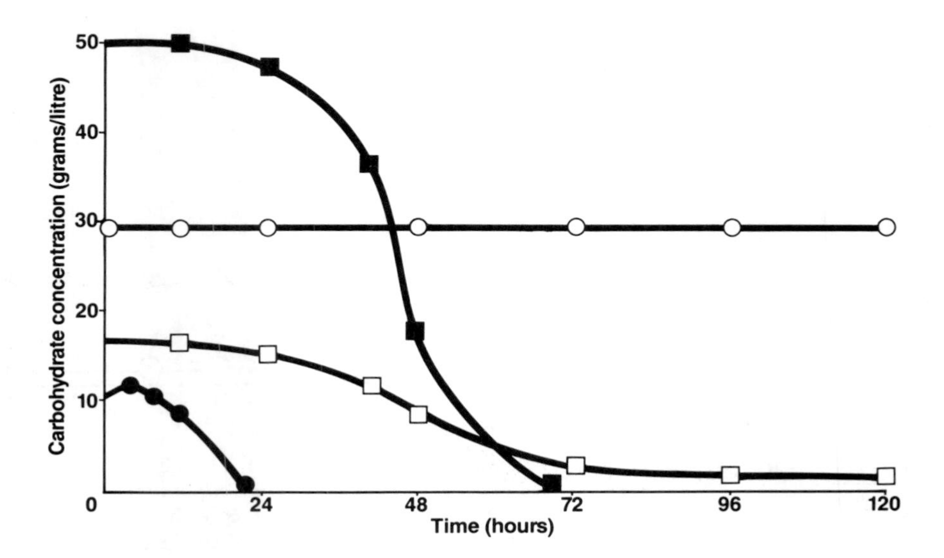

Figure 1. Uptake of the major wort sugars during fermentation by yeast: maltose (■), maltotriose (□), glucose (●) and dextrin (○).

The glucose analogue, 2-deoxyglucose (2-DOG), has been successfully employed by a number of workers (Bailey *et al.*, 1982; Laires *et al.*, 1983; Zimmerman and Scheel, 1977) for the selective isolation of spontaneous mutants of yeasts and other fungi. These mutants were derepressed for the production of carbohydrate hydrolyzing enzymes. Employing this non-metabolizable glucose analogue, derepressed mutants of brewing and other industrial yeast strains have been isolated that are able to metabolize maltose and maltotriose in the presence of high concentrations of glucose. In addition, 2-DOG starch mutants of *Saccharomyces diastaticus* have been isolated that exhibit increased fermentation ability in brewer's wort, cassava and corn mash (Stewart, 1987b).

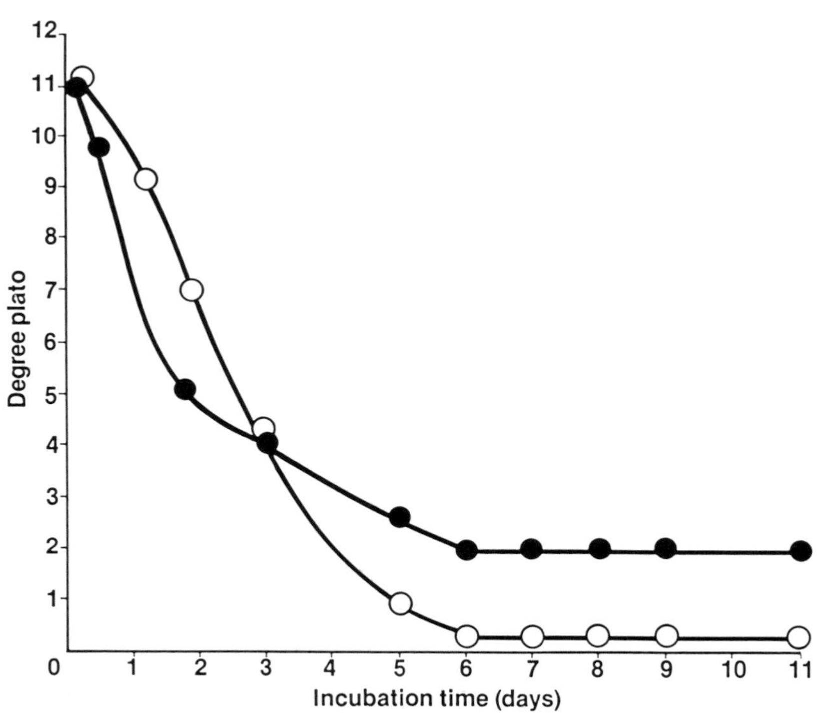

Figure 2. Static fermentation of an 11.3°P wort (40 litre scale) by a *Saccharomyces cerevisiae* brewing strain (●) and a *Saccharomyces diastaticus* diploid (*DEX1/DEX1, DEX2/DEX2, STA3/STA3*) (○).

In order to illustrate the phenomenon of derepression and the effect of 2-DOG mutants on overall fermentation rate of a brewer's wort, two brewing strains of *Saccharomyces cerevisiae* (coded 154 and 3001) have been studied (Stewart *et al.*, 1985b). In both instances (Figures 3 and 4), the presence of glucose repressed the uptake of maltose. A number of stable 2-DOG mutants were found to be capable of utilizing maltose in the presence of significant concentrations of glucose. For example, 2-DOG resistant mutants of *Saccharomyces cerevisiae* (strain 154) in a synthetic medium containing both maltose [8% (w/v)] and glucose [3% (w/v)] were able to completely metabolize the maltose (Figure 3A) whereas in the same medium, maltose uptake was slow with the parental strain and only 60% complete when fermentation ceased. Fermentation and ethanol formation rates in 12°P wort were also increased in the 2-DOG mutants when compared to the parental strain (Figure 3B).

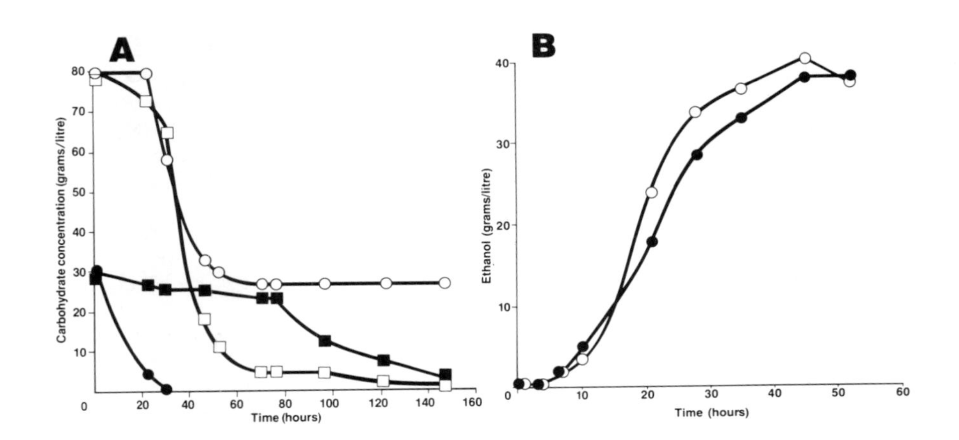

Figure 3. Fermentation characteristics of a brewing strain of *Saccharomyces cerevisiae* strain 154 and its derepressed maltose 2-DOG mutant. (A) Carbohydrate uptake in an 8% (w/v) maltose 3% (w/v) glucose-peptone yeast extract medium. Glucose uptake: parent (●) and mutant (■); maltose uptake: parent (○) and mutant. (B) Ethanol production in a 12°P wort: parent (●) and mutant (○).

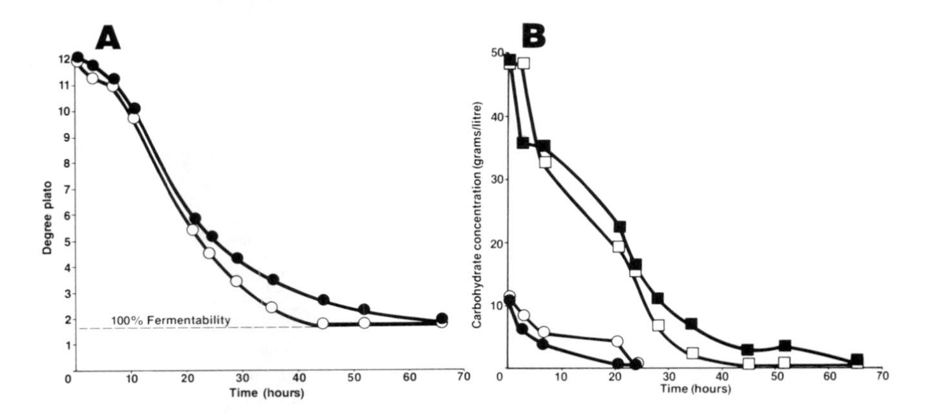

Figure 4. Fermentation characteristics of a brewing strain of *Saccharomyces cerevisiae* (strain 3001) and its derepressed 2-DOG mutant in a 12°P wort. (A) Fermentation rate: parent (●) and mutant (○); (B) Glucose uptake: parent (●) and mutant (○); maltose uptake: parent (■) and mutant (□).

Similar results were obtained with another strain of *Saccharomyces cerevisiae* (strain 3001) where maltose uptake in the 2-DOG mutants was not repressed by glucose. In 12°P wort, the overall fermentation rate was significantly faster in the 2-DOG mutants (Figure 4A), with complete fermentation being achieved in 45 h compared to 65 h in the parental strain. This increased fermentation rate was due to an increased maltose uptake rate in the 2-DOG mutants when compared to the parental strain with glucose having little influence upon maltose uptake in the mutants (Figure 4B). A trained taste panel operating in the triangular mode determined that the beer produced with 2-DOG mutants of both strains (eg., 154 and 3001) was not significantly different from that produced using the parental strains; all beers were produced under similar pilot plant brewing conditions. The mechanism by which 2-DOG resistant mutants are derepressed is far from understood but such mutants have been found to possess elevated levels of hexokinase, invertase and α-glucosidase (Russell *et al.*, 1987). Bailey and Woodward (1984) have described a mutant allele (designated *grr*-1) in *Saccharomyces cerevisiae* that is characterized not only by 2-DOG resistance, but also by insensitivity to glucose repression for invertase, α-glucosidase and galactokinase, as well as the mitochondrial enzyme cytochrome oxidase. The molecular mechanism(s) by which catabolite repression is effected in yeast is still largely unknown. Many genes are involved in this phenomenon as the regulatory mechanisms must include sensory and signaling systems for monitoring glucose availability and regulatory proteins that effect changes in expression of a multitude of genes.

SPHEROPLAST (PROTOPLAST) FUSION

Techniques that show greatest potential and promise as aids in the genetic manipulation of industrial yeast strains are spheroplast (or protoplast) fusion (also called somatic fusion) and transformation (Stewart and Russell, 1986a). Both of these techniques have a total disregard for ploidy and mating type and, consequently, have great applicability to many industrial strains because of their polyploid/aneuploid nature and absence of mating type characteristics.

The first step in spheroplast fusion is the removal of the yeast cell wall with lytic enzymes such as extracts of snail gut or enzymes from various microorganisms. Removal of the yeast cell wall leaves only the membrane surrounding the cytoplasm - this is a spheroplast. Such structures are osmotically fragile and only remain intact if maintained in a medium of high osmotic pressure, usually 0.8-1.2 M sorbitol in buffer. After thorough washing, to remove traces of the spheroplasting enzyme, the spheroplasts are suspended in the fusing agent which consists of polyethylene glycol (PEG) and calcium ions in buffer, and then are mixed with spheroplasts from a yeast strain with different genetic characteristics (Figure 5).

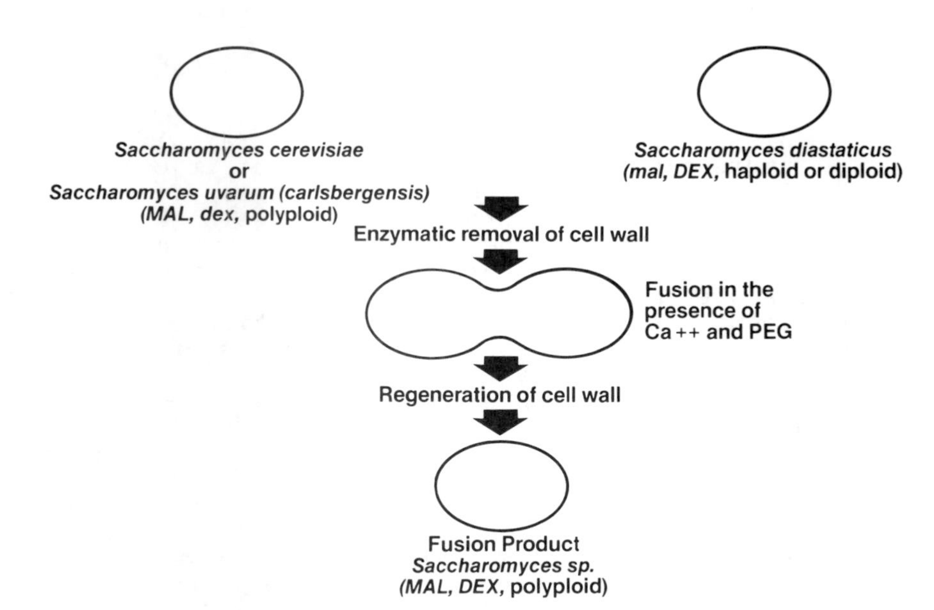

Figure 5. Procedure for spheroplast (protoplast) fusion in yeast.

After fusion (interchange of DNA material), the fused spheroplasts must be induced to regenerate their cell walls [this is achieved by embedding the spheroplasts in solid nutrient media containing 3% (w/v) agar and sorbitol] and to begin cell division. The mechanism of action of PEG as the fusing agent is not very well understood. It is believed to act as a polycation which together with calcium ions induces the formation of small aggregates of spheroplasts thus facilitating the interchange of DNA material (Jones *et al.,* 1987).

Using classical hybridization techniques, a diploid strain of the diastatic yeast species *Saccharomyces diastaticus* (coded 1384) was constructed with the following genotype: *a/α*, *DEX1/DEX1, DEX2/DEX2, STA3/STA3, mal/mal [DEX* and *STA* genes code for the production and secretion of glucoamylase and yeast strains that contain these genes in their genotype are able to utilize starch and dextrin (Stewart *et al.,* 1983)]. This strain has been fused with a brewing lager strain of *Saccharomyces uvarum (carlsbergensis)*. A successful fusion product has been isolated and coded 1400 in the Labatt Culture Collection and, in order to be patented (Stewart *et al.,* 1986a), has been deposited in the National Collection of Yeast Cultures, U.K. (coded NCYC 1460).

This particular fusion product (strain 1400) has interesting characteristics in that it can withstand elevated osmotic pressures, ferment at increased temperatures, and produce higher alcohol yields than all other *Saccharomyces* strains tested to date in this laboratory (Stewart *et al.,* 1984; D'Amore *et al.,* 1987). For example, in the production of industrial fermentation ethanol, particularly in the tropics, the use of a thermotolerant yeast strain, thus avoiding significant fermenter cooling, has a decided advantage. The fermentation characteristics of the fusion product (strain 1400) and the two fusion partners were investigated in a 20% (w/v) glucose medium at 40°C (Figure 6) (Stewart *et al.,* 1984). The *Saccharomyces uvarum (carlsbergensis)* strain fermented very poorly at this temperature with very little glucose uptake. The other fusion partner, a strain of *Saccharomyces diastaticus*, produced 6% (w/v) ethanol in 24 h. However, the fusion product (1400) proved to be the most thermotolerant of the yeast strains studied because it produced 7% (w/v) ethanol in 24 h and a final concentration of 8% (w/v) ethanol in 60 h of fermentation with 80% of the initial glucose being utilized.

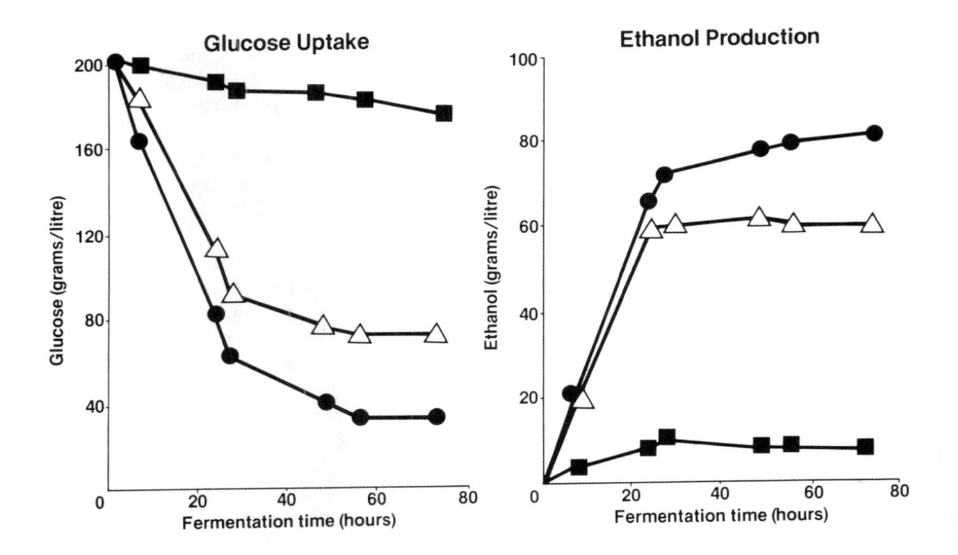

Figure 6. Fermentation characteristics of *Saccharomyces spp.* strains in PYN media with 20% (w/v) glucose at 40 C. *Saccharomyces uvarum (carlsbergensis)* lager brewing strain (3021) (■), *Saccharomyces diastaticus* strain (1384) (*DEX1/DEX1, DEX2/DEX2, STA3/STA3, malo/malo* (△), *Saccharomyces sp.* fusion product (1400) (●).

DIASTATIC YEASTS FOR DISTILLED ETHANOL PRODUCTION

The fermentation of starches to ethanol by yeasts requires pretreatment of the substrate in order to produce fermentable sugars. This pretreatment consists of three steps: gelatinization, liquefaction and saccharification. Gelatinization requires heat and free water and must precede liquefaction. Liquefaction, the dispersion of starch molecules into an aqueous solution, is accomplished by the use of heat and amylolytic enzymes. Heat stable α-amylases or malt enzymes may be employed. During liquefaction, starch molecules are only partially hydrolyzed producing a form of carbohydrate that cannot be assimilated by ethanol-producing yeasts such as *Saccharomyces cerevisiae*. Therefore, the partially hydrolyzed starch molecules must be converted to lower molecular weight sugars such as glucose and maltose by a process known as saccharification. This may be accomplished enzymatically, usually by the addition of fungal glucoamylase to the fermentation vessel at the time of yeast inoculation. The saccharifying glucoamylases represent a significant fraction of the total cost of producing ethanol via fermentation. Reduction of the amount of added glucoamylase could significantly decrease the cost of the final product.

It has been found possible to decrease glucoamylase addition to a starch mash fermentation by employing yeast which actively produce and secrete glucoamylase eg., strains of *Saccharomyces diastaticus* (Whitney *et al.*, 1985; Russell *et al.*, 1986). The fused genetically manipulated strain of *Saccharomyces*

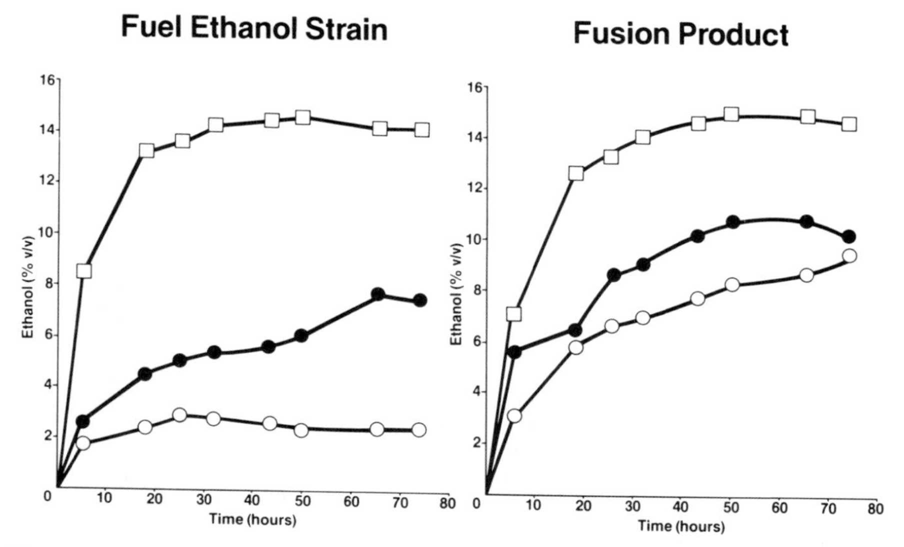

Figure 7. Ethanol production in a cassava mash with varying amounts of added glucoamylase: 0.00% (○); 0.01% (●); 0.10% (□).

diastaticus (strain 1400), discussed previously in this document, was studied as the glucoamylase-producing strain and compared to a strain of *Saccharomyces cerevisiae* (glucoamylase negative), that is currently being employed commercially in the production of fuel ethanol. The fermentation performance of the two strains was compared in a cassava mash with and without added glucoamylase. When the *Saccharomyces diastaticus* strain was employed, as a result of the secretion of glucoamylase due to the presence of *DEX* genes, added glucoamylase could be significantly decreased without reducing ethanol production or sugar uptake (Figure 7).

FLOCCULATION

As previously discussed, the flocculation property, or conversely lack of flocculation, of a particular yeast culture is one of the major factors when considering important characteristics during brewing and other ethanol fermentations. Unfortunately, a certain degree of confusion has arisen by the use of the term "flocculation" in the scientific literature to describe different phenomena in yeast cell behaviour. Specifically, flocculation, as it applies to brewer's yeast, is "*the phenomenon wherein yeast cells adhere in clumps and either sediment from the medium in which they are suspended or rise to the medium's surface*". This definition excludes other forms of aggregation, particularly those of "clumpy growth" and "chain formation" (Figure 8) (Stewart, 1975). This non-segregation of daughter and mother cells during growth has sometimes erroneously been referred to as flocculation. The term "non-flocculation" therefore applies to the lack of cell aggregation and, consequently, a much slower separation of (dispersed) yeast cells from the liquid medium. Flocculation occurs in the absence of cell division and only under rather circumscribed environmental conditions involving the cross-bridging of divalent cations, usually calcium, bridging anionic groups at the cell surface. Although yeast separation very often occurs by sedimentation, it may also be by flotation because of cell aggregates entrapping bubbles of carbon dioxide, as in the case of "top-fermenting" ale brewing yeast strains.

Individual strains of brewer's (and other

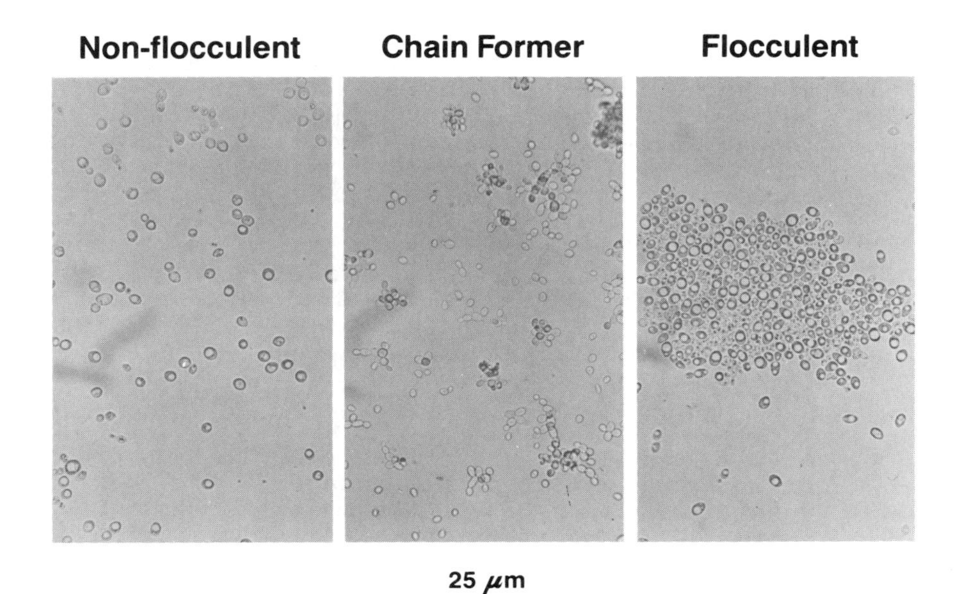

Figure 8. Light photomicrographs of non-flocculent, chain formers and flocculent yeast strains.

industrial) yeast differ considerably in flocculating power. At one extreme there are highly non-flocculent, often referred to as powdery, strains. At the other extreme there are flocculent strains. The latter tend to separate early from suspension in fermenting wort, giving a sweeter and less fully fermented beer. Poorly flocculent (non-flocculent or powdery) yeasts by contrast produce a dry, fully fermented, more biologically stable beer in which clarification is slow leading to the possible acquisition of yeasty flavours. The disadvantages presented by these two types of yeast are especially relevant to more traditional fermentation systems where the fermentation process is very dependent upon the sedimentation characteristics of the yeast. Contemporary brewing technology has largely reversed this situation where yeast sedimentation characteristics are now fitted into the fermenter design. The efficiency, economy and speed of batch fermentations have been improved by the use of cylindro-conical fermentation vessels and centrifuges (which are often employed in tandem).

There is little doubt that differences in the flocculation characteristics of various yeast cultures are primarily a manifestation of the yeast culture's cell wall structure. Studies in many laboratories have failed to reveal any meaningful differences in gross composition between the walls of the two culture types that would be directly correlated with the phenomenon of flocculation. Several mechanisms for flocculation have been proposed (Beavan *et al.*, 1979; Calleja, 1984). One hypothesis is that anionic groups of cell wall components are linked by Ca^{++} ions (Mill, 1964; Taylor and Orton, 1975); these anionic groups, in all likelihood are proteins (Calleja, 1974). Another hypothesis implicates mannoproteins specific to flocculent strains acting in a lectin-like manner to cross-link the cells; here Ca^{++} ions act as ligands to promote flocculence by conformational changes (Miki *et al.*, 1982).

Electron microscopy of flocculent and non-flocculent cultures shadowed with tungsten oxide has revealed that flocculent cultures possess a "hairy" outer surface (Figure 9) (Day *et al.*, 1975). It is worthy of note that surface appendages have been implicated in a number of instances of microbial flocculation, aggregation and adhesion (eg., Achtman, 1975; Calleja *et al.*, 1975; Douglas, 1985; Ottaw, 1975). For example, it is believed that adhesion of *Candida albicans* cells to mucosal surfaces appears to involve lectin-like interactions between the protein portion of mannoprotein located in fibrils on the yeast surface and glycoside receptors on epithelial cells (Figure 10). It would, therefore, appear that flocculation of brewer's yeast strains and the adhesion of *Candida albicans* to mucosal surfaces have many features in common (Stewart and Russell, 1986b).

Confirmation that yeast flocculation is genetically controlled came from the pioneering research of Gilliland (1951) and Thorne (1951). However, because of the polyploid/aneuploid nature of brewing yeast strains most, but not all, (Johnston and Reader, 1983; Stewart and Russell, 1977, 1981)

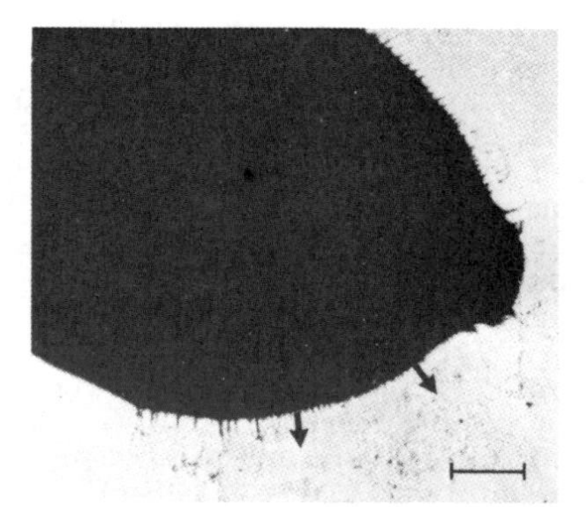

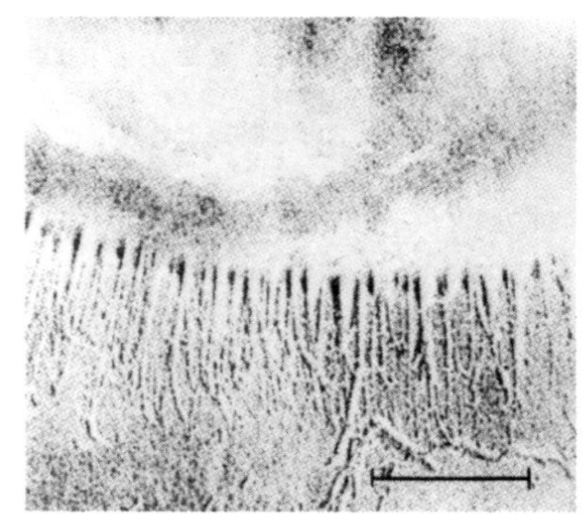

Flocculent Yeast

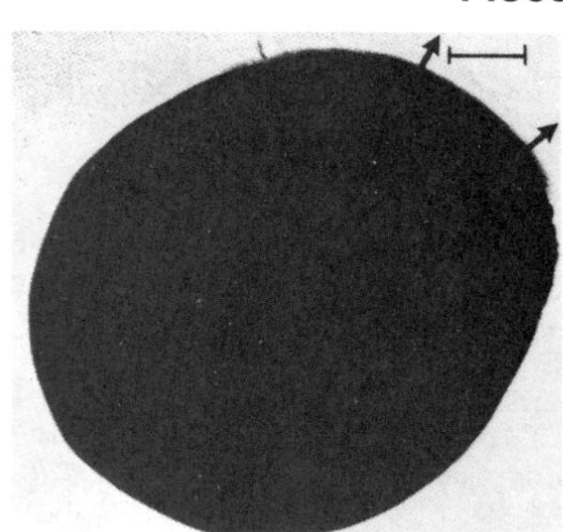

Non-flocculent Yeast

[a]Shadow-cast with tungsten oxide. Bar = 1μm.

Figure 9. Electron photomicrographs of flocculent and non-flocculent strains of *Saccharomyces cerevisiae* shadow cast with tungsten oxide. Bar = 1 μm.

of the research on flocculation genetics has been conducted on haploid/diploid genetically defined laboratory strains. A number of genes have been reported to directly influence the flocculence phenotype in *Saccharomyces spp.*: *FLO1* (Stewart and Russell, 1977), *flo3* (Lewis *et al.*, 1976), *FLO5* (Hodgson *et al.*, 1985), *FLO8* (Yamashita and Fukui, 1983), *tup 1* (Tipke and Hull-Pillsbury, 1984) and *fsu1* (Holmberg and Kielland-Brandt, 1978). However, at this time the role of these genes is far from understood. As flocculation is a property associated with the structure of cell walls and particularly with wall proteins, a study of the genetics of yeast flocculation affords an opportunity to study the genetics of structural

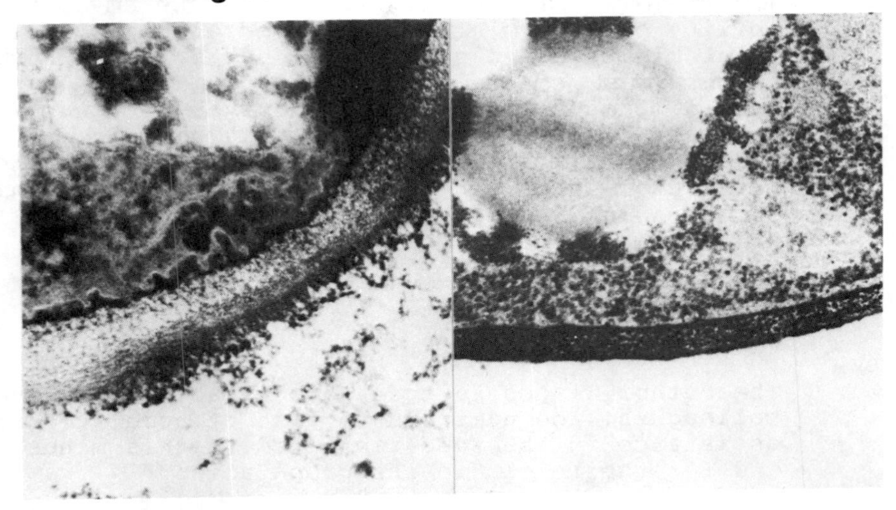

Figure 10. Electron photomicrographs of adhering and non-adhering cultures of *Candida albicans*. Photographs courtesy of Dr. L.J. Douglas, University of Glasgow, Scotland.

rather than enzymatic proteins. This research also presents the possibility of being able to control and manipulate one of the most important characteristics of a brewer's yeast strain.

CONCLUSIONS

It is important that ethanol producing yeast strains be able to ferment a wide range of sugars and carbohydrates, tolerate the prevailing environmental conditions and possess the required flocculation characteristics.

With the advent of novel genetic manipulation methods such as spheroplast fusion and transformation, it is worth asking the question "what is the 'ideal' ethanol-producing microorganism?" It should possess the broad spectrum sugar and carbohydrate utilizing properties and ethanol and thermotolerance of *Saccharomyces diastaticus*, together with the amylolytic ability of *Schwanniomyces castellii* and also produce a palatable beverage similar to that from *Saccharomyces cerevisiae* and *Saccharomyces uvarum (carlsbergensis)*. It should possess the cellulase activity of *Clostridium thermocellum*, the osmotolerance of *Saccharomyces rouxii*, the pentose utilizing ability of *Candida shehatae*, the lactose utilizing ability of *Kluyveromyces fragilis* and the high specific fermentation rate and near theoretical ethanol yield of *Zymomonas mobilis*.

ACKNOWLEDGEMENTS

The author wishes to thank his Labatt Research colleagues, too numerous to list by name, for their assistance in the research cited in this manuscript and for their help in its preparation.

REFERENCES

Achtman, M. 1975. Mating aggregates in *Eschericha coli* conjugation. *J. Bacteriol. 123*: 505-515.

Bailey, R.B. and A. Woodward. 1984. Isolation and characterization of a pleiotropic glucose repression resistant mutant of *Saccharomyces cerevisiae. Mol. Gen. Genet. 193:* 507-516.

Bailey, R.B., T. Benitez and A. Woodward. 1982. *Saccharomyces cerevisiae* mutants resistant to catabolite repression, use in cheese whey hydrolysate fermentation. *Appl. Environ. Microbiol. 44*: 631-636.

Beavan, M.J., D.M. Belk, G.G. Stewart and A.H. Rose. 1979. Changes in electrophoretic mobility and lytic enzyme activity associated with development of flocculating ability in *Saccharomyces cerevisiae. Can. J. Microbiol. 25*: 888-895.

Bilinski, C.A., I. Russell and G.G. Stewart. 1986. Analysis of sporulation in brewer's yeast: induction of tetrad formation. *J. Inst. Brewing 92:* 594-598.

Bilinski, C.A., I. Russell and G.G. Stewart. 1987a. Physiological requirements of the induction of sporulation in lager yeast. *J. Inst. Brewing 93*: 216-219.

Bilinski, C.A., I. Russell and G.G. Stewart. 1987b. Cross-breeding of *Saccharomyces cerevisiae* and *Saccharomyces uvarum (carlsbergensis)* by mating of meiotic segregants: isolation and characterization of species hybrids. In: *Proceedings of the 21st Congress of the European Brewery Convention*, pp. 497 - 504, IRL Press Limited, Oxford, U.K.

Calleja, G.B. 1974. On the nature of the forces involved in sex-directed flocculation of a fission yeast. *Can. J. Microbiol. 20*: 797-803.

Calleja, G.B. 1984. *Microbiol Aggregation* CRC Press, Inc. Boca Raton, FL., U.S.A.

Calleja, G.B., B.Y. Yoo and B.F.Johnson. 1977. Fusion and errosion of cell walls during conjugation in the fission yeast *Schizosaccharomyces pombe*. J. Cell Sci. 75: *139–146*.

D'Amore, T., C.J. Panchal, I. Russell and G.G. Stewart. 1988. Osmotic pressure effects and intracellular accumulation of ethanol in yeast during fermentation. J. Indust. Microbiol. 2: *365–372*.

Day, A.W., N.J. Poon and G.G. Stewart. 1975. Fungal fimbriae. III. The effect of flocculation in Saccharomyces cerevisiae. Can. J. Microbiol. 21: *558–564*.

Douglas, L.J. 1985. Adhesion of pathogenic Candida *species to host surfaces.* Microbiol. Sci. 2: *243–247*.

Gilliland, R.B. 1951. The flocculation characteristics of brewing yeast during fermentation. In: Proceedings of the 3rd Congress of the European Brewery Convention, *pp. 35–52*.

Hodgson, J.A., D.R. Berry and J.R. Johnston. 1985. Discrimination by heat and proteinase treatments between flocculent phenotypes conferred on Saccharomyces cerevisiae *by genes* FLO1 *and* FLO5. J. Gen. Microbiol. 131: *3219–3227*.

Holmberg, S. and M.C. Kielland-Brandt. 1978. A mutant of *Saccharomyces cerevisiae* temperature sensitive for flocculation. Influence of oxygen and respiratory-deficiency on flocculence. *Carlsberg Res.Commun. 43*: 37-47.

Johnston, J.R. and H.P. Reader. 1983. Genetic control of flocculation. In: *Yeast Genetics. Fundamental and Applied Aspects* (Spencer, J.F.T., D.M. Spencer and A.R.W. Smith eds.), pp. 205-224, Springer-Verlag, New York, U.S.A.

Jones, R.M., I. Russell and G.G. Stewart. 1987. Classical genetic and protoplast fusion techniques in yeast. In: *Yeast Biotechnology* (Berry, D.R., I. Russell and G.G. Stewart eds.), pp. 53-79. Allen and Unwin, London, U.K.

Laires, A., I. Spencer-Martins and N. van Uden. 1983. Use of D-glucosamine and 2-deoxyglucose in the selective isolation of mutants of the yeast *Lipomyces starkeyi* derepressed for the production of extracellular endodextrins. *Z. Allg. Mikrobiol. 13*: 601-612.

Lewis, C.W., J.R. Johnston and P.A. Martin. 1976. The genetics of yeast flocculation. *J. Inst. Brewing 82*: 158-160.

Lipke, D.M. and C. Hull-Pillsbury. 1984. Flocculation of *Saccharomyces cerevisiae tup1* mutants. *J. Bacteriol. 159*: 797-799.

Lyons, T.P. 1981. *Fuel Alcohol: A Step to Energy Independence*. Alltech Technical Publications, Lexington, KY., U.S.A.

Miki, B.L., N.H. Poon, A.A. James and V.L. Seligy. 1982. Possible mechanism for flocculation interaction governed by gene *FLO1* in *Saccharomyces cerevisiae. J. Bacteriol. 150*: 878-889.

Mill, P.J. 1964. The nature of the interactions between flocculent cells in the flocculation of *Saccharomyces cerevisiae. J. Gen. Microbiol. 35*: 61-68.

Ottaw, J.G.C. 1975. Ecology, physiology and genetics of fimbriae and pili. *Annu. Revs. Microbiol. 29*: 79-96.

Russell, I., C.M. Crumplen, R.M. Jones and G.G Stewart. 1986. Efficiency of genetically engineered yeast in the production of ethanol from dextrinized cassava starch. *Biotech. Letters 8*: 169-174.

Russell, I., R.M. Jones, D.R. Berry and G.G. Stewart. 1987. Isolation and characterization of derepressed mutant of *Saccharomyces cerevisiae* and *Saccharomyces diastaticus*. In: *Biological Research on Industrial Yeasts*. Vol. II (Stewart, G.G., I. Russell, R.D. Klein and R.R. Hiebsch eds.) pp. 57-65, CRC Press, Inc., Boca Raton, FL., U.S.A.

Sage. R. 1987. The policy background to oxygenated gasolines. In: *Sixth Canadian Bioenergy R&D Seminar* (Granger, C. ed.) pp. 43-44, Elsevier Applied Science, London, U.K.

Stewart, G.G. 1975. Yeast flocculation - practical implications and experimental findings. *Brew. Dig. 50(3)*: 42-47.

Stewart, G.G. 1981. The genetic manipulation of industrial yeast strains. *Can. J. Microbiol. 27*: 973-990.

Stewart, G.G. 1987a. Twenty five years of yeast research. *Develop. Indust. Microbiol. 29:* - in press.

Stewart, G.G. 1987b. Yeast - the primary industrial microorganism. In: *Biological Research on Industrial Yeasts*. Vol. I (Stewart, G.G., I. Russell, R.D. Klein and R.R. Hiebsch eds.), pp. 1-20, CRC Press, Inc., Boca Raton, FL., U.S.A.

Stewart, G.G. and I. Russell. 1977. The identification, characterization and mapping of a gene for flocculation in *Saccharomyces sp. Can. J. Microbiol. 24*: 441-447.

Stewart, G.G. and I. Russell, 1981. Yeast flocculation. In: *Brewing Science*. Vol. II (Pollock, J.R.A. ed.), pp. 61-92, Academic Press, London, U.K.

Stewart, G.G., C. Panchal, I. Russell and A.M. Sills. 1983a. Advances in ethanol from sugars to starch - a panoramic paper. In: *Ethanol from Biomass* (Duckworth, H.E., and E.A. Thompson eds.) pp. 4-52. Royal Society of Canada, Ottawa, Canada.

Stewart, G.G., C.J. Panchal and I. Russell, 1983b. Current developments in the genetic manipulation of brewing yeast strains - a review. *J. Inst. Brewing 89*: 170-188.

Stewart, G.G., C.J. Panchal, I. Russell and A.M. Sills. 1984. Biology of ethanol-producing microorganisms. *CRC Crit. Revs. Biotech. 1*: 161-188.

Stewart, G.G., C.A. Bilinski, C.J. Panchal, I. Russell and A.M. Sills. 1985a. The genetic manipulation of brewer's yeast strains. In: *Microbiology 1985* (Schlessinger, D. ed.) pp. 367-374. ASM Publications, Washington, D.C., U.S.A.

Stewart, G.G., R.M. Jones and I. Russell, 1985b. The use of derepressed yeast mutants in the fermentation of brewery wort. In: *Proceedings of the 20th Congress of the European Brewery Convention*, pp. 243-247, IRL Press Limited, Oxford, U.K.

Stewart, G.G. and I. Russell. 1986a. One hundred yeast of yeast research and development in the brewing industry. *J. Inst. Brewing 92*: 537-558.

Stewart, G.G. and I. Russell, 1986b. The relevance of the flocculation properties of yeast in today's brewing. In: *E.B.C. – Symposium on Brewer's Yeast.* Monograph - XII, pp. 53-70. Verlag Hans Carl, Nurnberg, F.R.G.

Stewart, G.G., I. Russell and C.J. Panchal. 1986. Genetically stable allopolyploid somatic fusion product useful in the production of fuel alcohols. Canadian Patent No. 1,199,593.

Taylor, N.W. and Orton, W.L. 1975. Calcium in flocculence of *Saccharomyces cerevisiae*. *J. Inst. Brewing 81*: 53-57.

Thorne, R.S.W. 1951. Some aspects of yeast flocculation. In: *Proceedings of the 3rd Congress of the European Brewery Convention*, pp. 21-34.

Whitney, G.K., C.R. Murray, I. Russell and G.G. Stewart. 1985. Potential cost savings for fuel ethanol production by employing a novel hybrid yeast strain. *Biotech. Letters 7:* 364-354

Yamashita, I. and S. Fukui. 1983. Mating signals control expression of both starch fermentation genes and a novel flocculation gene *FLO8* in the yeast *Saccharomyces*. *Agric. Biol. Chem.* *47*: 2889-2896.

Zimmerman, F.K. and I. Scheel. 1977. Mutants of *Saccharomyces cerevisiae* resistant to carbon catabolite repression. *Mol. Gen. Genet.* *155*: 75-83.

26

Manipulation of flavour production by yeast: physiological and genetic approaches

David R. Berry
Department of Bioscience and Biotechnology, University of Strathclyde,
204 George Street, Glasgow, G1 1XW, Scotland

Although the distilling industry involves just as much fermentation as the brewing or wine industries the it has traditionally placed much greater emphasis on the distillation and maturation parts of the process when it comes to controlling the flavour. Although there can be little doubt that changes do occur during these processes it is a mistake to minimise the role of the fermentation since it is at this stage that the majority of the flavour compounds are produced. In the case of gin or vodka production the distillation stage effectively eliminates the organoleptic compounds produced during the fermentation so it is reasonable to use the cheapest form of carbohydrate substrate and to use a yeast and fermentation process to maximise yield with no regard to flavour. However in the production of whiskies and brandies, organoleptic compounds produced in the fermentation remain in the product and make a contribution to its flavour. Presumably process parameters which influence the numbers and concentrations of these compounds also influence the organoleptic properties of the final product.

In the Scottish malt whisky industry malted barley is the only carbohydrate source which can be used in the process. Since this is a very expensive form of carbohydrate it is reasonable to assume that the industry considers that this makes a significant contribution to the final flavour. Part of this contribution can clearly be attributed to the peaty flavour of peated malt but to acquire this would not necessitate the use of 100% barley. It is generally considered in the Scotch whisky industry that malt whisky has a much better flavour than the maize derived cheaper grain spirit. This is distilled in the lowlands and blended with malt whisky during the production of blended whisky which is the major product. It is reasonable therefore to ask, what contribution the barley makes to the flavour of the final product. Is it essential or is it just a historical anachronism which could be replaced by sugar syrups if the law permitted? Looking at brandy production, similar questions can be posed. Is the difference between two brandies influenced by the quality or origin of the wine used or is it purely dependent upon the distillation technique adopted?

It is our experience that the nature of the barley is important in influencing spirit quality. However, there is little information in the literature to justify such a belief. Over several years of carrying out small scale whisky fermentations in the laboratory and analysing the spirits obtained from them, we noticed that a considerable amount of the variability observed in these experiments could be attributed to the use of different batches of barley. (see Table 1, Ramsay 1982). This suggests that there are some features of the barley which influence flavour production and which can vary from batch to batch.

The next question to be posed is which of the components of barley are most important in influencing flavour. Malting and mashing are complex processes which give rise to a complex substrate, wort, which contains a mixture of sugars, amino acids and peptides, trace elements, growth factors and some lipids. The relationship between higher alcohol production and amino acid metabolism has been extensively studied (MacDonald et al 1984, Watson & Berry 1987); however there is not a simple relationship between availability of a particular amino acid and the production of the corresponding higher alcohol since higher alcohols can be produced by an anabolic route as well as the classical catabolic route. However if a large excess of an amino acid is provided, an effect on higher alcohol production is observed (Table 2, Ramsay 1982). It has been reported by several groups working in the fields of brewing, (Engan 1981), wine production, (Kunkee & Goswell 1977) and distilling, (Berry & Ramsay 1983) that higher alcohol production varies with yeast strain thus indicating that it should be subject to genetic manipulation. The isolation of mutants which produce higher levels of isoamyl acetate in sake has also been reported by Ashida et al. (1987).

In fact the known relationship between higher alcohol production and the biosynthesis and catabolism of amino acids offers scope for the genetic manipulation of higher alcohol formation. The biosynthetic pathways leading to most amino acids in yeast have been well worked out (Jones & Fink 1982) so it is possible to predict which enzymes and genes are important in controlling the level of production of a

Table 1: Variation in concentration of higher alcohols, fatty acids and esters formed in worts prepared from a single and different batches of grist

mg/l	Single Batch of Grist (n=8)					Different batches of grist (n=14)				
	min	max	mean	S.D.	CV%	min	max	mean	S.D.	CV%
Propanol	16.1	20.0	18.4	1.7	9.0	16.1	25.3	19.4	4.4	12.2
iso-Butanol	52.2	58.2	54.7	2.00	3.7	52.2	80.0	64.3	11.2	17.3
Amyl alcohols	119	126	123	3.1	2.5	119	192	145	26.6	18.3
2-Phenyl-ethanol	49.9	57.5	53.0	2.2	4.2	49.9	75.3	61.2	9.9	16.2
Total	239	258	249	6.8	2.7	239	363	290	48.4	16.7
Octanoic acid	3.2	4.6	3.9	0.4	12.6	3.2	6.6	4.7	1.2	24.6
Decanoic acid	1.5	2.4	1.9	0.37	18.7	1.5	3.9	2.5	0.87	33.5
Dodecanoic acid	0.23	0.34	0.26	0.4	14.2	0.23	0.81	0.41	0.20	48.8
Ethyl acetate	47.5	54.2	49.5	2.2	4.4	47.5	66.1	53.6	5.8	10.8
iso-Butyl acetate	0.26	0.36	0.31	0.03	10.1	0.26	0.39	0.34	0.04	11.4
Amyl acetate	4.3	4.8	4.6	0.16	3.6	4.3	6.3	5.1	0.69	13.3
2-Phenyl-ethyl acetate	1.9	2.3	2.1	0.16	7.7	1.9	3.5	2.4	0.52	21.4

Table 2: Effect of 50% substitution of wort with a solution of amino acids on higher alcohol formation (mg/l)

Amino acids	No addition	Glucose only	Leucine (2.12)	valine (1.88)	Threo-nine (1.91)	phenyl alamine (2.65)
Propanol	17.6	6.1	18.6	46.9	20.2	71.8
Isobutanol	56.0	34.8	27.3	435	41.2	44.2
Amyl alcohols	124	98.4	502	80.6	286	70
2-Phenyl ethanol	53.6	37.5	14.2	16.1	28.0	643.6
n-butanol	-	-	-	1.38	1.75	-

Amino acid solution values given are in g/l, in 12.5% glucose solution. The values correspond to 225mg/l amino nitrogen.

Table 3: Concentration of esters produced in continuous fermentation of an 80g/l Glucose medium compared with a batch fermentation of wort (SG 1.053)*

Ester	Continuous Fermentation of glucose (mg/l distillate)	Batch Fermentation of wort (mg/l distillate)
Ethyl acetate	514	499
Amyl acetate	27	71
Ethyl hexanoate	5	6
Ethyl octanoate	17	25
Ethyl decanoate	12	23
Ethyl dodecanoate	1.5	16

*From Berry & Chamberlain (1986).

given higher alcohol. Wine yeasts which produce reduced levels of higher alcohols have been produced by classical genetic techniques (Rous et al 1983). A strain which lacked -isopropylamate dehydratase was isolated by selecting for leucine auxotrophs. When this was used in wine fermentations the wine produced was found to contain 20% less total higher alcohol and 50% less isoamyl alcohol. It should not be difficult to obtain similar results with distilling yeasts if this were desirable.

Another group of flavour compounds, the vicinal diketides, are also produced from the amino acid biosynthetic pathways and could be manipulated in a similar manner. Diacetyl and 2,3-pentadione are produced from the biosynthetic pathways leading to isoleucine and valine. The genes for most of the enzymes involved in these biosynthetic pathways are known, and those for threonine deaminase, aceto hydroxy acid reductoisomerase, dihydroxyacid dehydratase and a branched chain amino acid aminotransferase have been cloned. (Holmberg 1984, Casey 1986).

One wort component which we believe is important in flavour development is the lipid or trub. Saccharomyces cerevisiae has a requirement for unsaturated fatty acid and sterol for growth. These can be synthesised by the yeast in aerobic or even microaerophilic conditions but not in strictly anaerobic conditions. If growth is occurring then this requirement must be being satisfied either because there is sufficient oxygen for the cell to synthesise its own, because of reserves in the cell or because they are present in the medium. Again this requirement has been studied in detail in brewing because of the use of anaerobically grown inocula and the need for raising fermentations. Since most distilling processes use aerobically grown yeasts which contain a high level of unsaturated fatty acids and sterols the need to study this area of metabolism has not been apparent. However, it is possible that an excess of these compounds either in the yeast inoculum or derived from the wort can repress ester formation. We have demonstrated that adding back the fine solid fraction from unfiltered wort to a clear wort does repress ester formation (Ramsay & Berry 1983). This lipid fraction should be absent from good clear wort however changes in mashing procedure could allow some lipid to pass through. Studies in continuous culture have indicated that in strictly anaerobic conditions no growth occurs but that if sufficient oxygen is present to permit growth then the level of ester production can be low. This could be a difficult parameter to control in any continuous distilling process. In these experiments, in which defined medium was used, great difficulty was experienced in producing a spirit with an ester composition which was comparable to a batch fermentation. This was found to be attributable to the low level of glucose in a carbon limited continuous culture. When this was increased by doubling the inlet glucose concentration then a pattern of esters resembling that found in a batch fermentation was obtained (Table 3. Berry & Chamberlain 198). This change in the concentration of esters at higher glucose concentration was found to be associated with a dramatic change in the ratio of the concentration of ester to parent alcohol indicating that the change cannot be attributed to a change in the availability of alcohols (Table 4, (Chamberlain, 1985). The levels of esters obtained in the spirit produced under high glucose conditions are of the same order as those observed in a simple batch fermentation. These results correlate well with the reports that the use of high gravity worts in brewing leads to increased ester production. Conversely if a low level of esters was required it should be possible to achieve this by either aerating (raising) the fermentation or adding the majority of the wort as a feed (a fed batch process) at such a rate that the level of glucose in the wash remains low.

Most of the genetic manipulation which has been carried out with brewing and distilling yeast has been aimed at improving the yield of alcohol obtained from starch derived substrates. Most of the genetic techniques which are available for manipulating the yeast genotype have been applied to brewing and distilling yeasts in recent years (Young 1986). Most strains have been obtained by a simple screening procedure from the mixed populations of yeast which were present in traditional processes. The application of hybridisation techniques was limited because many of the traditional brewing strains were polyploid or aneuploid and not amenable to classical hybridisation. However brewing and distilling strains with an amyloglucosidase activity have been obtained using classical and rare mating techniques (Tubb 1983, Stewart & Russell 1987). It has been reported that the amount of glucoamylase which was required for a corn mash in a distilling process could be reduced when a *S. diastaticus* strain was used which has a 1,4-glucoamylase activity (Russell & Stewart 1986).

Genetic techniques can also be used to eliminate off-flavours. Crosses between *S. diasataticus* and brewing strains of *S. cerevisiae* produced strains which gave phenolic off-flavours when used in brewing. This was found to be attributable to the presence of an enzyme which decarboxylates ferulic acid present in the wort to 4 vinyl guaiacol. This is then reduced to 4 ethyl guaiacol. The gene for this enzyme had been introduced into the new strains with the glucoamylase gene, however it was subsequently possible to eliminate it by further genetic manipulation. (Russell 1984, Stewart & Russell 1986).

Mutation and hybridisation techniques are probably easier to apply to the elimination of a specific flavour or off-flavour (see above) rather than to the introduction of an improved flavour. This is because the overall flavour of alcoholic beverages is dependent upon the action of several hundred genes such that random elimination or recombination of these genes is more likely to have a deleterious effect on flavour

Table 4: Ratios of acetate esters to parent alcohols at different glucose concentrations

Glucose concentration g/l	Ethyl acetate / Ethanol x10	Amyl acetate / Amyl alcohol x10
40	1.3	<1.0
60	16.3	8.2
80	16.1	8.6
100	19.2	12.8
120	15.9	9.5

Table 5: Concentration of acetic acid and ethyl acetate in distillates from wash adjusted to various pH values with or without the addition of acetic acid

Addition of acetic acid g/l	pH	Acetic acid mg/1 in distillate	Ethyl acetate mg/1 in distillate
NIL	2.0	N.D.	128
	3.5	N.D.	131
	4.3(CONTROL)	N.D.	130
	5.5	N.D.	132
1.05	2.0	397	169
	3.5	399	139
	4.3	232	137
	5.5	30.4	130
2.10	2.0	1079	184
	3.5	923	150
	4.3	689	147
	5.5	450	140

N.D. - None detected

rather than to improve it. Mutants which affect sulphur metabolism may be useful in limiting the production of sulphur compounds such as H S,dimethyl sulphide and S-methyl thioacetate which can produce off-flavours in alcoholic beverages including spirits (Nykanen & Suomalainen 1983). However many of these are already present from the wort.
Temperature is a critical parameter in distillery fermentations. Elevated temperatures can lead to loss of volatiles, especially esters, increased ethanol toxicity and a decrease in yeast viability (Tsuji & Kamaguchi (1984) have reported that less isoamyl alcohol is accumulated at 30 C than at 28 C because more is lost by evaporation. We have demonstrated in small scale whisky fermentations that yeast autolysis caused by elevated temperatures can result in an increase in the level of contamination by lactobacilli. One consequence of this is the increased level of lactic acid and the consequent decrease in the pH of the wash (Lavery _et al_ 1986). However it seems unlikely that the pH achieved is low enough to cause any significant increase in the amount of esters produced during the distillation phase since previous experiments carried out in our laboratory had indicated that this only occurs when washes with a pH of 2-2.5 are distilled (Table).

REFERENCES

Ashida, S., Ichikawa, E., Suginama, K. & Imagasu, S. (1987) Isolation and application of mutants producing sufficient isoamyl acetate, a sake flavour component. Agric. Biol. Chem. 51, 2061-65.

Berry, D.R. (1984) Physiology and microbiology of the malt whisky fermentation. In Progress in Industrial Microbiology, 19, Ed M.E. Bushell, 199-244, Elsevier, Amsterdam.

Berry, D.R. & Ramsay, C.M. (1983) The whisky fermentation, past, present and future. In Current Developments in Brewing and Distilling, Ed F.G. Priest and I Campbell. 45-58.

Berry, D.R. & Chamberlain, H. (1986) Formation of organoleptic compounds by yeast grown in continuous culture on a defined medium. J. Amer.Soc. Brew. Chem. 44, 52-56.

Berry, D.R. & Watson, D.C. (1987) Production of organoleptic compounds, In Yeast Biotechnology, Ed D.R. Berry,. I. Russell & Stewart. -, George Allen & Unwin.

Casey, G.P. (1986) Cloning and analysis of two alleles of the ILV3 gene from _Saccharomyces carlesbergenis_. Carlesberg. Res. Commun. 51, 327-41.

Engan, S. (1981) Beer Composition, volatile substances. In Brewing Sciences, vol. 2. Ed J.R.A. Pollock, 98-165. Academic Press, London.

Holmberg, S. (1984) Genetic improvements of brewers yeast. In Trends in Biotechnology, 2, 98-102.

Jones, R., Russell, I. & Stewart, G.G. (1987) Classical genetic and protoplast fusion techniques. In Yeast Biotechnology, Ed D.R. Berry, I. Russell & G.G. Stewart, - George Allen & Unwin. 55-79.

Jones, F.W. & Fink, G.R. (1982) Regulation of amino acid and nucleotide biosynthesis in yeast. In. The Molecular Biology of The Yeast Saccharomyces. Ed. Strathern, J.N., Jones, E.W. and Broach, J.R. 2, 181-299.

Kunkee, R.E. & Goswell, R. (1977) Table wines, In Economic Microbiology. Alcoholic Beverages. Ed A.H. Rose, 314-386.

Lavery, M., Chamberlain, H. & Berry, D.R. (1986) Use of a small scale insulated fermentation to simulate the Scotch Malt Whisky fermentation. Food Microbiology 3, 157-159.

MacDonald, J., Reeve, P.T.V., Ruddlesden, J.D. & White, F.H. (1984) In Progress in Industrial Microbiology, 19, 47-198, Elsevier, Amsterdam.

Nyakenen, L. & Suomalainen, H. (1983) Aroma of Beer, Wines and Distilled Alcoholic Beverages. D. Reidel Publishing Company, Dortrecht, Netherlands.

Ramsay, C.M. (1982) PhD Thesis. University of Strathclyde

Russell, I. (1984) MSc Thesis. University of Strathclyde.

Stewart, G.G. & Russell, I. (1986) One hundred years of yeast research and development in the brewing industry. J. Inst. Brew. 92(6), 539-558.

Stewart, G.G. & Russell, I. (1987) In Yeast Biotechnology, Ed D. R. Berry, I. Russell and G.G. Stewart. - , George Allen & Unwin.

Tsugi, K. & Kanaguchi, N. (1986) Formation of isoamyl acetate during fermentation of distillers wort. J. Soc. Brewing (Japan) 78 547-55.

Tubb, R.S. (1983) Genetic pathways to super-attenuating yeasts. In. Current Developments in Malting, Brewing and Distilling. Ed F.G. Priest & I. Campbell. 67-82.

27

Selection and breeding of yeast suitable for high temperature fermentation

K. Takahashi and K. Yoshizawa
National Research Institute of Brewing, 2-6-30 Takinogawa, Kita-ku, Tokyo 114, Japan

ABSTRACT

One yeast strain (1031R) with a high fermentation rate at 40 °C was selected from 706 strains collected in Thailand, and identified as *Saccharomyces cerevisiae* by taxonomic studies. The specific growth rate (μ) and mean generation time (t_d) of this yeast at 40.5 °C were 0.480 hr^{-1} and 1.44 hr respectively. The μ value was larger than that of *sake* yeast *Saccharomyces cerevisiae* Kyokai no. 7 (K-7) which was used as control (μ, 0.139 hr^{-1}; t_d, 4.99 hr). This yeast produced an acid off-flavour in mash (but this flavour did not transfer to distillates). In order to improve this off-flavour, protoplast fusion between 1031-R2 (adenine auxotrophic mutant of 1031R) and high flavour compound producing *sake* yeast Kyokai no. 7 was carried out. Five stable fusants (two fusants were tetraploid) were isolated, but strains that produced ethanol rapidly at 40 °C with little off-flavour were not found among them. Ninety-six spores from AM2, one of the tetraploid fusants, were isolated with a micro-manipulator, and two diploid segregants were chosen. The protoplast fusion and segregation method used were useful for breeding of a new brewing yeast.

INTRODUCTION

The production of alcoholic beverages by fermentation has gained considerable attention recently, a part of which has been focused on lowering the production cost. In the production of alcoholic beverages, the main mash is generally cooled to a suitable temperature for yeast fermentation. Therefore, the use of yeast having a high temperature tolerance is desirable for reducing cooling energy, and also to shorten the fermentation period. We report here the selection and use of yeast suitable for high temperature fermentation, and also breeding of a new yeast for flavour improvement [1-3].

MATERIALS AND METHODS

1. Selection of yeast suitable for high temperature fermentation

(1) Yeast strains

For selection of yeast suitable for high temperature fermentation, 706 strains collected in Thailand were used, and *sake* yeast *Saccharomyces cerevisiae* Kyokai no. 7 (K-7) was used as a control. This yeast is popularly used for the making of *sake* which is traditional Japanese rice wine.

(2) Yeast selection method

(a) First selection: One hundred µl of preincubated yeasts was inoculated in 3 ml of YM-ethanol medium (0.3% yeast extract, 0.3% malt extract, 0.5% peptone, 2% glucose and 5% ethanol, medium A) or in 3 ml of YM medium (only ethanol is absent from medium A, medium B). Yeasts in medium A and medium B were incubated for five days at 35 °C and 40 °C respectively, and yeasts that grew quickly to high density in both media were selected.

(b) Second and third selection: Fifty µl of preincubated yeasts was inoculated in 40 ml of *koji*-extract medium (21.3% glucose) and incubated for 8 days at 35 °C (second selection), and yeasts selected with second selection were incubated in the same medium for 8 days at 40 °C (third selection). Yeasts having greater ethanol-producing capability than K-7 were selected.

(3) Determination of growth curve at various temperatures

Yeasts were preincubated in 5 ml of *koji*-extract medium (10.5% glucose) for 2 days at 30 °C, and harvested. Yeast cells (2×10^7) were inoculated in 10 ml of *koji*-extract medium (21.3% glucose), and cultured with shaking. The yeast growth was followed by absorbance at 660 nm automatically.

(4) Ethanol producing capability of yeast

Yeast cells ($2x10^6$ ml^{-1}) were incubated in *koji*-extract medium (16.2% glucose) for 6 days at 40 °C, and ethanol content in the medium was determined daily by gas chromatography.

(5) Ethanol tolerance of yeast

Different concentrations of ethanol (0 - 10%) were added to YM medium containing 5% glucose, and yeast cells ($2x10^6$ ml^{-1}) were incubated for 7 days at 35 °C or 40 °C. Growth was measured by absorbance at 660 nm.

2. Fermentation of corn grits without cooking

Cells (10^9) of 1031R strain were incubated in 1 l of 0.1 M succinic acid buffer (pH 4.5) containing 200 g of uncooked corn grits (size: 200 - 800 µm) and commercial saccharifying enzyme (0.2 g). Fermentation was carried out at 40 °C for several days.

3. Breeding a new brewing yeast

(1) Yeast strains

The yeast strains used were alcohol-thermotolerant strain 1031R-2 and high flavour producing *sake* yeast K-7. The 1031R-2 was adenine auxotrophic mutant (ade/ade) of 1031R obtained by the procedure of Ouchi et al. [4] and formed ascospore (spo^+). *Sake* yeast K-7 showed temperature sensitive growth on the ß-alanine medium (ßal/ßal) in which pantothenate was replaced with ß-alanine, and did not form ascospore (spo^-). These strains were all diploid.

(2) Medium

The complete medium for growth was YPAD which contained 1% yeast extract, 2% peptone, 0.04% adenine sulphate and 2% glucose. The defined minimal medium was ß-alanine medium containing 1 M sorbitol. ß-Alanine medium contained $(NH_4)_2SO_4$ 0.05%, KH_2PO_4 0.15%, $MgSO_4.7H_2O$ 0.05%, citric acid 0.016%, glucose 2%, thiamine-HCl 200 ppb, pyridoxine-HCl 200ppb, nicotinic acid 200 ppb, inositol 100 ppb, p-aminobenzoic acid 200 ppb, biotin 0.2 ppb, ß-alanine 40 ppb and agar 3%, and its pH was adjusted to 5.0.

(3) Protoplast fusion between 1031R-2 strain and K-7 strain

The yeast cells of each strain cultured for 18 hr at 30 °C in YPAD medium were harvested, washed twice with SP buffer (1 M sorbitol, 0.1 M phosphate buffer, pH 7.0), incubated for 30 min at 30 °C in SP buffer containing 0.2% 2-mercaptoethanol and 60 mM EDTA, and protoplasted with 0.24 mg ml^{-1} Zymolyase 5000 (Seikagaku Kogyo Co., Tokyo) in SP buffer containing 0.2% of 2-mercaptoethanol for 3 hr at 30 °C. These protoplasts were washed twice with SP buffer containing 10 mM $CaCl_2$, and the protoplasts were poured onto the selective agar medium (ß-alanine medium) containing 1 M sorbitol, overlaid the selective agar medium and incubated for 7 - 10 days at 37 °C.

(4) Segregation of spores

Spores obtained from protoplast fusant were segregated using a micro-manipulator.

(5) Determination of DNA content

Nucleic acid was extracted by the method of Herbert et al. [5], and DNA content in extracts was determined by the method of Burton et al. [6].

(6) Evaluation of off-flavour

The intensity of off-flavour of fermented mashes with corn grits without cooking was evaluated organoleptically on a 3 point scale from 1 (weak) to 3 (strong).

4. Ethanol determination by gas chromatography

Sample solution (0.5 ml), 10 ml of distilled water and 1 ml of 2% (v/v) acetone (as an internal standard) were mixed for ethanol determination. The conditions of analysis were: column, glass (1 m x 3 mm) packed with 15% PEG-600 on Gasport A; column temperature, 75 °C; injection temperature, 120 °C; carrier gas, N_2 (40 ml min^{-1}); sample volume, 0.5 µl.

RESULTS AND DISCUSSION

Selection of yeast suitable for high temperature fermentation

1. Selection of yeast

Yeast selection was based on ability to grow and to ferment at 40 °C. From 706 strains, we obtained 124 strains in the first selection, and in the second selection we obtained 5 strains. Finally, 2 strains (1031R and 94-2) were selected in the third selection. These 2 strains were identified as *Saccharomyces cerevisiae* by taxonomic studies [7].

2. Characteristics of selected yeasts

Growth curves of selected yeasts at 40 °C are shown in Figure 1, and specific growth rate (µ) and mean generation time (t_d) at various temperatures are shown in Table 1. The two selected yeasts began to grow after a shorter lag time and attained a larger growth value than K-7, and of the two strains, 1031R strain showed more tolerance to high temperature than 94-2 strain. Generally, ethanol-producing yeast *Saccharomyces cerevisiae* has the largest µ value at 30 - 36 °C as shown in *sake* yeast K-7, but 1031R, a

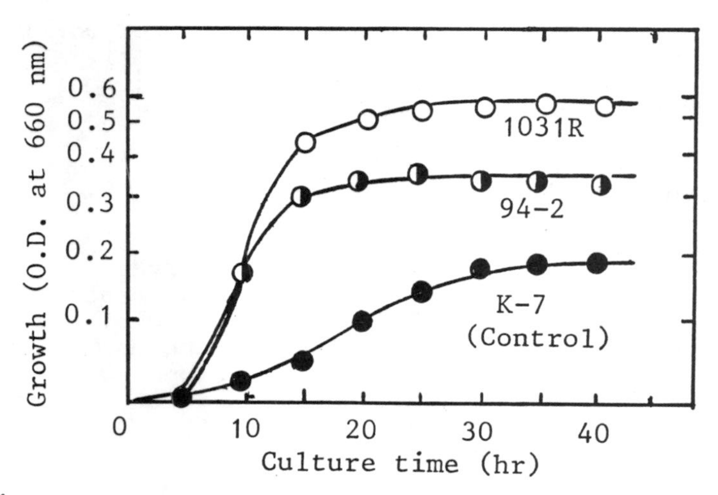

Figure 1. Growth curve of selected yeasts at 40°C. Medium, *Koji*-extract (21.3% glucose)

Table 1. Specific growth rate and mean generation time of selected yeasts at various temperatures.

Temp. (°C)	K-7		1031R		94-2	
	μ	t_d	μ	t_d	μ	t_d
26	0.433	1.60	0.433	1.60	0.445	1.56
30	0.476	1.46	0.456	1.52	0.446	1.55
33.5	0.548	1.26	0.462	1.50	0.442	1.57
36	0.511	1.36	0.462	1.50	0.442	1.57
38	0.447	1.55	0.488	1.42	0.436	1.59
40.5	0.139	4.99	0.480	1.44	0.396	1.75

μ, specific growth rate (/hr); t_d, mean generation time (hr). Medium, *Koji*-extract (21.3% glucose)

new isolate, showed the largest μ value at 38 - 40 °C. It seemed characteristic of the properties of the thermotolerant yeast that the growth rate did not differ greatly with incubation temperature. The time courses of ethanol production of 1031R and K-7 strains were compared. 1031R strain produced 44.6 g l^{-1} ethanol in 3 days fermentation at 40 °C; a value more than twice that of the K-7 strain. In 6 days fermentation at 40 °C, 1031R strain and K-7 strain produced 53 g l^{-1} ethanol and 25 g l^{-1} respectively, and the difference between the two strains was increased. The growth of K-7 strain was inhibited by the increase of ethanol at 35 °C, but that of 1031R strain was not inhibited until the ethanol concentration reached 10% (v/v). In the case of incubation at 40 °C, neither yeast could grow in medium which contained 4% (v/v) ethanol.

Utilization of selected yeast for ethanol fermentation of corn grits without cooking

1. Selection of saccharifying enzyme

As an appropriate saccharifying enzyme for uncooked corn grits, amylase A prepared from *Aspergillus niger* (crude) was selected from 14 commercial enzymes. Enzyme activities of amylase A were as follows: α-amylase (a), 13,300 (U g^{-1}); glucoamylase (g), 67,320 (mg glucose hr^{-1} g^{-1}); acid protease (p), 2,500 (μg Tyr hr^{-1} g^{-1}); the ratio of a:g:p = 0.2:1:0.04. Optimum temperature and pH of amylase A were 45 °C and 3.5 respectively, and these values were lower in comparison with other commercial saccharifying enzymes. Therefore, this enzyme was though to be suitable for ethanol fermentation using selected yeast 1031R at 40 °C.

2. Ethanol fermentation of uncooked corn grits

In ethanol fermentation of raw starchy materials, the fermentation rate was affected by rate of glucose production, so a considerable amount of saccharifying enzyme is used at present. We have studied the ethanol fermentation of uncooked corn grits by using only 0.1% of enzyme to substrate. The time courses of ethanol fermentation under various conditions are shown in Figure 2. The most effective ethanol fermentation with raw corn grits was to stir about 100 rpm and to use minerals (K, 300 ppm; Mg and P, 200 ppm) and cellulase (0.1% to corn grits). Enzyme action was accelerated and yeast population increased by the addition of significant amounts of these minerals. Using this procedure, ethanol concentration attained 58.9 g l^{-1} in 140 hr fermentation, and this ethanol yield corresponded to 85% of theoretical ethanol concentration based on raw starch content.

Breeding a new brewing yeast for improvement of off-flavour

The selected yeast, 1031R, produced an acid-like off-flavour in fermented mash, but this flavour did not transfer to distillates and so is not a practical problem. We have bred a new brewing yeast which could produce ethanol rapidly at 40 °C with little off-flavour using this yeast and *sake* yeast K-7. *Sake* yeast K-7 produces large amounts of higher alcohols and esters.

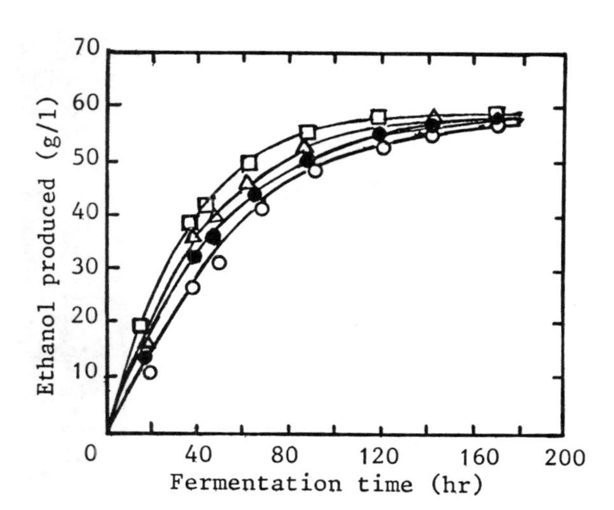

Figure 2. Effects of agitation and minerals on ethanol fermentation with corn grits without cooking. Yeast, 1031R; Fermentation temp., 40°C
○, not stirring; ●, stirring; △, stirring (minerals were added); □, stirring (minerals and cellulase were added)

1. Protoplast fusion between 1031R strain and sake *yeast K-7*

Breeding a new brewing yeast was carried out as shown in Figure 3. Protoplast fusion was performed between 1031R-2 and K-7 several times, and 5 stable fusants, AM1 - AM5, were obtained. All these fusants were prototrophs, and also formed ascospores in the sporulation medium (1% potassium acetate, 0.1% yeast extract and 0.05% glucose). As shown in Table 2, AM2 and AM4 were larger in cell size and cell volume, approximately twice each parent, and the amounts of DNA in them were about the sum of that in the parental strains. From these results, AM2 and AM4 were likely to be tetraploid fusants and they might be obtained by fusion in 1 : 1 ratio of each parental strain. Among 5 fusants, strains which produced ethanol rapidly at 40 °C with little off-flavour were not found.

2. Spore segregation by tetrad analysis

We expected to isolate interesting strains by spore segregation of protoplast fusants when spore segregation from AM2, one of tetraploid fusants, was carried out. Of 96 spores isolated from AM2, 55 germinated. The K-7 strain had few spores, and the germination of spores from 1031R-2 was lower than from AM2 (sporulation rate of 1031R-2 was 25%). The examination of fermentative ability of 55 segregants

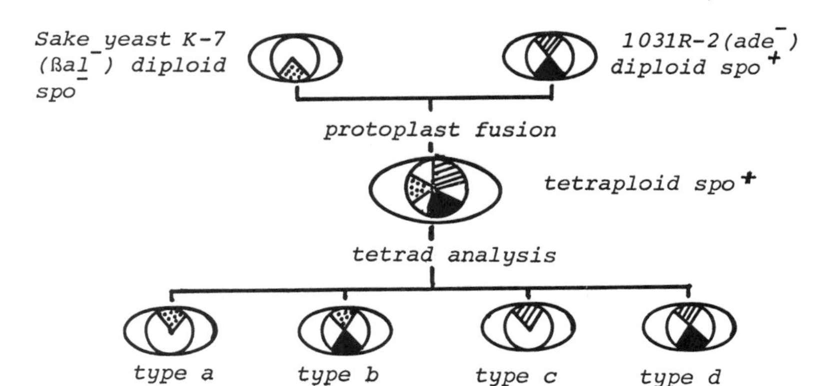

Figure 3. Schematic illustration for breeding of a new brewing yeast suitable for high-temperature fermentation. ■, genes for fast production of ethanol at 40°C. ▨, genes for production of off-flavour. ⁙, genes for production of good flavour. Our purpose was to breed a type *b* strain.

Table 2. Cell size, DNA content and ethanol production of fusants and sensory evaluation of off-flavour of their fermented mashes.

Strain	Cell Size (μm)	Cell Volume (μm^3)	DNA content (μg/10^8cells)	Ethanol* concn. (g/l)	Intensity** of off-flavour
K-7	6.7 x 5.0	88	6.5	26.0	1.0
1031R-2	5.7 x 4.9	72	5.1	51.3	3.0
AM1	6.1 x 5.2	86	5.7	53.7	3.0
AM2	8.5 x 5.1	193	10.7	30.8	2.0
AM3	7.5 x 4.5	133	8.1	36.3	3.0
AM4	7.7 x 5.4	168	11.7	38.7	3.0
AM5	7.2 x 5.2	141	9.6	52.9	3.0

* 10^9cells were incubated in 100 ml of corn grits suspension (corn grits 20 g, amylase A 0.04 g, 7.5% lactic acid 0.46 ml, distilled water 100 ml, pH 4.5), and fermented for 4 days at 40°C. ** Intensity of off-flavour was assessed on a three-point scale from 1 (weak) to 3 (strong).

showed that they produced between 15.8 g l^{-1} and 51 g l^{-1} ethanol, and 20 segregants were selected as strains producing more than 40 g l^{-1} ethanol in YM medium (10% glucose). As shown in Table 3, 5 segregants showed high ethanol producing capability and reduced off-flavour, especially AM2-17B and

Table 3. Ethanol production of selected segregants from AM2 and sensory evaluation of off-flavour of the fermented mashes.

Strain	Ethanol concn. (g/1)	Intensity of off-flavour	Strain	Ethanol concn. (g/1)	Intensity of off-flavour
K-7	26.0	1	AM2-11D	47.3	2.2
1031R-2	51.3	2.6	AM2-17B	47.3	1.8
			AM2-18C	47.3	1.7
AM2- 3B	49.7	2.0	AM2-22C	50.5	2.2

Ethanol fermentation and sensory evaluation were the same as shown in Table 2. These fusants were all prototrophs except AM2-17B strain ($ßal^-$).

AM2-18C which were close to our object. The protoplast fusion and spore segregation method we used was useful for the breeding of a new brewing yeast.

REFERENCES

1. Takahashi, K., Ogata, S., Yoshizawa, K., Nakamura, K., Karuwanna, P., Kumunuanta, J. J. Brew. Soc. Japan **81**, 124, 1986.
2. Nagai, H., Takahashi, K., Yoshizawa, K. J. Brew. Soc. Japan **81**, 268, 1986.
3. Miyazaki, S., Kitamoto, K., Takahashi, K., Yoshizawa, K. Hakkokogaku Kaishi **65**, 1, 1987.
4. Ouchi, K., Shimoda, M., Nakamura, Y., Kojima, Y., Nishiya, T. Hakkokogaku Kaishi **61**, 349, 1983.
5. Herbert, D., Phipps, P.J., Strange, R.E. Methods in Microbiology 5B, 324, Academic Press, New York, 1971.
6. Burton, K. Biochem J. **62**, 315, 1956.
7. Lodder, J. (ed) The Yeasts: a Taxonomic Study 2nd ed., North-Holland Publ., 1970

28

Supercritical fluids for extract preparation

D. A. Moyler
Felton Worldwide Ltd., Bilton Road,
Bletchley, Milton Keynes, MK1 1HD, UK

ABSTRACT

The selective solvent characteristics of high pressure CO_2, and the low temperature at which extraction is carried out, make it a versatile extractant for the labile aromatic flavour botanicals used for manufacture of distilled alcoholic beverages. This paper reviews the properties and uses of CO_2, with and without entrainers, giving a detailed description of the commercial equipment used for the liquid CO_2 extraction of natural botanical materials. Features of the composition of extracts prepared using this equipment are compared with oils made by steam distillation, using organoleptic profiles and GLC techniques. Specific examples of botanicals used in gin and liqueur manufacture are discussed, which indicate that the CO_2 extracts are closer in composition to the oils as they occur in nature than are those obtained by other techniques.

INTRODUCTION

The constraints placed upon extracts used in alcoholic beverages fit for human consumption are many. Legislation, or worse, possible future legislation, E numbers and the subsequent consumer preference for 'naturals', leads the extraction industry to entrench with the traditional water and ethanol [1,2] and investigate the technological, high pressure solvents. Of these, carbon dioxide is without doubt the safest and most versatile practical extractant [3,4].

The selective solvent characteristics of high pressure CO_2, and the low temperature at which extraction is carried out, make it a selective extractant for the labile aromatic flavour botanicals used in the manufacture of distilled alcoholic beverages. This paper reviews the properties and uses of CO_2, with and without entrainers, giving a detailed description of the commercial equipment used for the liquid CO_2 extraction of natural botanical materials.

REVIEW OF THE USE OF CO_2

I will deal with the uses of CO_2 in its broadest context of operation, in the subcritical and supercritical modes. Subcritical conditions are below the critical point in the phase diagram of CO_2 [5]; that is 31 °C and 74 bar. Practical subcritical conditions of operation as an extraction solvent are between 50 and 80 bar, and 0 and 10 °C. Under these conditions, liquid CO_2 has been used commercially for 6 years to selectively extract essential oils. Under the supercritical conditions of fluid CO_2, between 200 and 300 bar, and 40 to 50 °C, CO_2 has been used to extract hops and decaffeinate coffee since 1976 but is not currently used commercially to extract flavour oleoresins.

There is no doubt of the versatility and commercial practicality of the use of solid, liquid and gaseous CO_2 in the food and beverage industries [6]. However, to make a meaningful examination of the technical literature as it relates to the distilled alcoholic beverage industries, it is best to study detailed applications of the use of CO_2.

An alternative to distillation of spirits

Firstly, can CO_2 seriously be considered as a commercial alternative to the distillation of spirits? Certainly not for Cognac, Armagnac or whisky, where so many related trace components of the natural fermentation and maturation processes are integral to the flavour of the spirit. However, studies by Kuk and Montagna [7] and others have shown from the tertiary phase diagrams of water, ethanol and CO_2, that it is technically possible to remove the water from ethanol-water mixtures with liquid CO_2.

At conditions near the critical point, ethanol and

low molecular weight organic molecules transfer to the lighter CO_2 phase. Water is removed from the sump of the system, the pressure released and the CO_2 gas reclaimed, reliquified, and reused. The ethanol is then removed from the sump, and the whole cycle can be repeated. The concept was developed in the USA by Critical Fluid Systems Inc., part of Arthur D. Little Inc., who claim that only one third of the energy is needed, compared to distillation, "more than balancing the additional capital costs".

The original patent, granted in 1979 to de Filippi and Vivian [8] of Critical Fluid Systems, was applied to the removal of water from alcohols, as well as technical solvents. In 1983 they claimed to be "past pilot scale and into commercial scale soon". With such a system it would be feasible to make vodka, schnapps and gin. Attempts to make whisky are likely to give a novel beverage rather than the true flavour. This patent was not intended for use in the alcoholic beverage industry and will not be a serious competitor to distillation techniques. It probably forms part of a blanket patent coverage, to protect the company against all possible uses.

To recover alcohol from fermentation wastes

The US Department of energy has been developing the CO_2 technique to extract ethanol from ferments, any unfermented sugars or higher molecular weight materials being virtually insoluble in CO_2. The applications are likely to be for motor fuel in countries with readily available fermentation sources and no natural fuel oil, which are without the currency necessary to buy it, e.g. Brazil.

To concentrate flavours from ferments

A logical extension from alcohol recovery is to use a high quality fermentation source such as cider, and extract it with CO_2 under pressure to extract the alcohol with flavour, producing a Calvados or schnapps type of beverage. The extract, standardised with water, would be the completed product. Such extracts may find uses for low cost commercial alcoholic beverages, liqueurs and cocktail bases.

To prepare low alcohol beverages

Although not strictly the province of distilled beverages, it is worth mentioning the feasibility of using CO_2 to reduce the alcohol content of wines and beers. The process was studied by Berger et al. in 1981 [9]. Pressurised liquid CO_2 was used to remove the alcohol and flavour concentrate, and the CO_2 evaporated by depressurisation. The concentrate was fractionally distilled to remove alcohol, and the flavour added back to the beverage. The fractionation stage is particularly difficult in practice, and the commercial reality is that direct distillation of the beverage at atmospheric or reduced pressure, dialysis and reverse osmosis techniques are used to make low alcohol beers [10] and wines.

For extraction of fruit flavours

The high moisture content of crushed fruits presents physical problems for CO_2 extraction. Some notable studies, at the US Department of Agriculture in California by Schultz [11], and by several Russian workers, have worked on the problems in relation to apples, pears and vegetables (USA) and various fruit juices, honey, cocoa and tea (USSR). Crushed fruit pulps have to be partly clarified to remove bulk cellulose before liquid-liquid CO_2 extraction. Such pulps are best preserved because of the risk of microbiological spoilage. Sulphur dioxide is undesirable for preservation because of the bleaching of some fruits, particularly red fruits. Benzoate salts are preferred, especially as they have little solubility in CO_2, so giving clean, stable preservative-free extract.

An alternative technique is to clarify the fruit juice and evaporate some of the water. The concentrated, or folded juice as it is known, is then absorbed into an inert powder, and packed into steel pressure columns for solid-liquid CO_2 extraction. Ethanol is frequently used as an entrainer with the liquid CO_2, to modify its solvent polarity. These extracts do contain the full fruity lactone body from the fruit, but suffer from missing some of the green topnotes of the fresh fruit. They contain some extra caramel type notes, caused by caramelisation of natural fruit sugars during the original folding of the fruit juice. These jam-like notes can be used to advantage in some flavour applications.

The direct extraction of fruits in a solid-liquid CO_2 system is technically feasible but expensive. Low yields, high capital costs of equipment and transportation costs restrict the siting of CO_2 extraction equipment to the fruit growing areas.

The extracts have to compete commercially with both crushed fruits and their juices, for use in food and beverages. In most beverage applications, the colours and mouth feel of fruit juice sugars are desirable characteristics. The uses of low colour, sugarless, CO_2 extracts are limited to speciality products such as compounded clear colourless fruit vodkas, colourless liqueurs or as flavour boosters to juice based liqueurs.

For extraction of herbs and spices

Pioneering laboratory work on the use of liquid CO_2 as an extraction solvent was carried out at Krasnodar in Russia in the 1960's and 1970's. Conditions of operation between 10 and 20 °C and subcritical pressures were used. Since then, useful work has been accomplished in this field by, amongst others, Stahl and Schilz [12], Volbrecht [13], Calame and Steiner [14], Weldon [15] and Fincher [16].

Subcritical and supercritical CO_2 techniques have been investigated for the extraction of spices in particular. As a generalisation, subcritical extraction yields the undegraded essential oil, plus absolute components with molecular weights up to 400 [17]. Supercritical extraction yields concretes and oleoresins, with all the extractables, including those components of poor solubility in alcohol. It has been confirmed in practice, that the subcritical CO_2 extracts are more soluble in flavoured spirits and liqueurs and they are being used to develop flavour concentrates.

PROPERTIES OF CO_2

The physical properties of CO_2 at 20 °C are [18]:
* Liquid at pressures between 5 and 74 bar
* Low dipole moment, hence low polarity (comparable to pentane)
* Low viscosity, so penetrates (CO_2 0.07 cP, ethanol 1.2 cP, water 1.0 cP)
* Low latent heat of evaporation (CO_2 42 cal g^{-1}, ethanol 204 cal g^{-1}, water 540 cal g^{-1}). This low latent heat enables CO_2 to be recovered using little energy, especially as it can be boiled at 13 °C in a commercial system [19].
* Chemically pure and stable
* Colourless
* Odourless and tasteless
* Easily removed
* Selective

An example of a commercial CO_2 extraction plant is shown in Figure 1. Fuller details are given by Gardner [19]. There are four process variables: temperature, pressure, particle size and CO_2 throughput. Each can affect the extract and they all need to be optimised for each botanical. The judgement of optimisation should be based on the

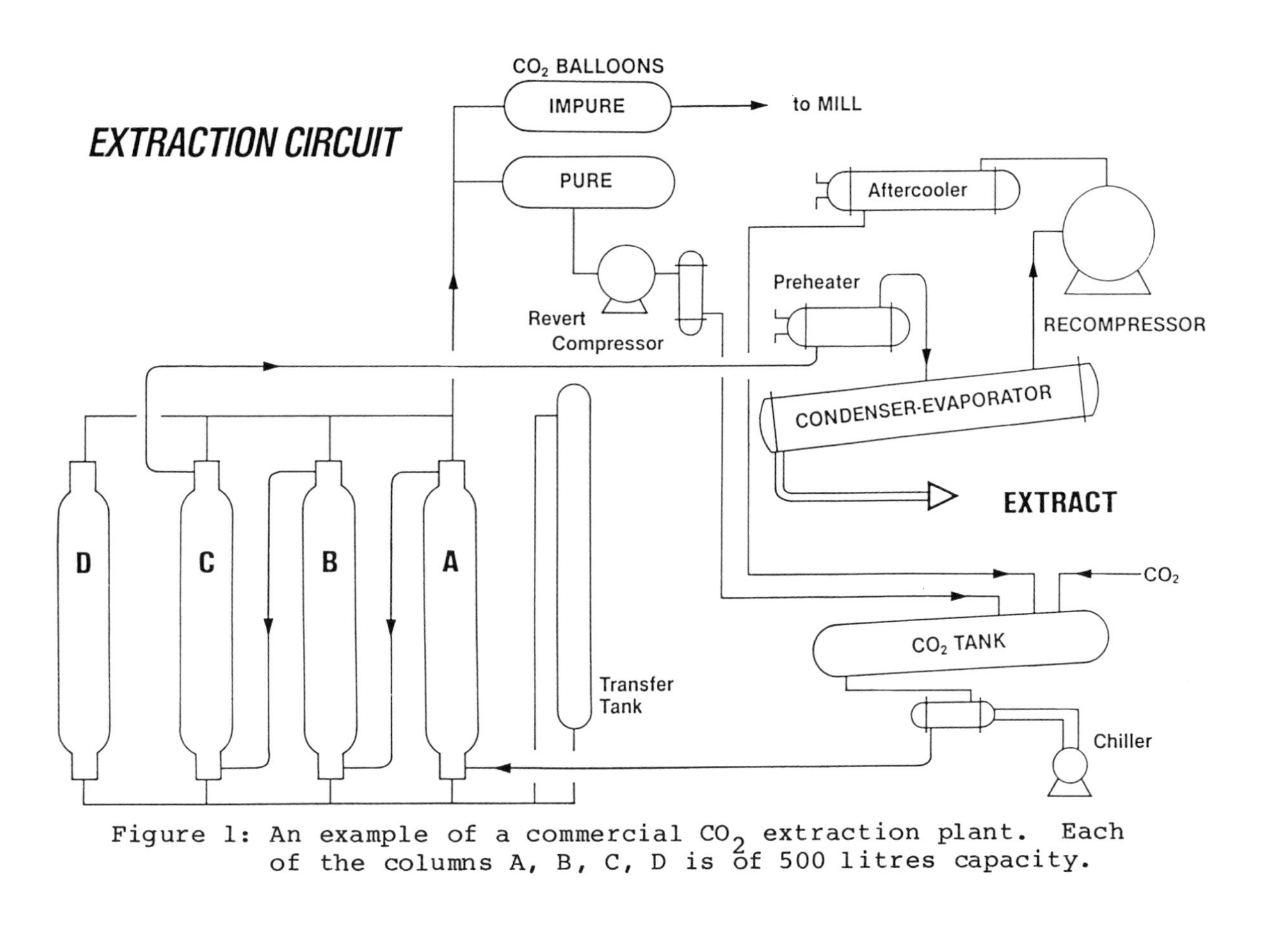

Figure 1: An example of a commercial CO_2 extraction plant. Each of the columns A, B, C, D is of 500 litres capacity.

flavour and odour of the extract, not yield. This is an example of the technique of solid-liquid CO_2 extraction. Other variants include the addition of metered amounts of ethanol entrained in the CO_2 flow. The effect of an entrainer is usually to increase the polarity of the solvent mixture and so modify the profile of the components extracted.

A further variant, liquid-liquid CO_2 extraction was described as a method of extracting the flavour of fruit juices. The partly clarified juice is partitioned by pumping against a counter-current flow of liquid CO_2. The flavour is more soluble in the CO_2, leaving the spent sugary flavourless juice. The CO_2 is evaporated, recovered and reused, the flavour concentrate being tapped from the sump of the system.

PROPERTIES OF LIQUID CO_2 EXTRACTED ESSENTIAL OILS

These oils exhibit some or all of the following six features:

* No solvent residues
* No off notes generated during distillation
* More topnotes because of low temperature extraction
* More backnotes because CO_2 extracts the character of an absolute
* Better solubility because no terpenes are generated during extraction
* Concentration of aromatic components

CO_2 EXTRACTS AND THEIR USES

Natural extracts which can be used to improve the flavour of some compounded alcoholic beverages are:

Juniper berry oil

A comparison of the organoleptic profiles of CO_2 extracted and steam distilled oils of juniper berry (*Juniperus communis*) shows the CO_2 oil to be smoother, fresher, less resinous, musty and terpeney than the steam distilled oil. All of these attributes are advantageous in gins and liqueurs. Gas chromatographic comparisons have already been published [20], which show that the CO_2 extract has a much lower level of monoterpenes, hence better solubility and more flavour.

The commercial CO_2 extract is an example of liquid CO_2 used with an ethanol entrainer. This has the dual effect of increasing the polar components extracted and at the same time diluting the oil to inhibit the polymerisation that can occur with this oil. The result is increased flavour, good solubility and stability.

Ginger oil

A comparison of the organoleptic profiles of CO_2 extracted and steam distilled oils of ginger (Zingiber officinale, Roscoe) is dramatic. The CO_2 extract is more sweetly aromatic, lemony, spicy, warm and above all, pungently hot. These characteristics are all important in spicy liqueurs and ginger wines. Published GLC comparisons [22-25] show more green topnotes and lower monoterpene hydrocarbon contents in the CO_2 extract. These low monoterpene levels enable the CO_2 oil to be dissolved directly into ginger ale beverages.

Clove bud oil

A comparison of the organoleptic profiles of CO_2 extracted and steam distilled oils of clove buds (Eugenia caryophyllata) shows the CO_2 oil to be warmer, fruitier, less phenolic and musty. These characteristics enable more of this oil to be used as a modifier, to give warmth to coffee liqueurs, without the medicinal phenolic harsh flavour which can be obtained with other clove oils.

GLC comparisons show [21] twice the eugenyl acetate level and correspondingly lower free eugenol content in the CO_2 oil of clove bud. If the CO_2 oil is steam distilled in the laboratory, the resulting oil is lower in eugenyl acetate content. This indicates that considerable degradation occurs during these wet heat distillation conditions of processing.

Nutmeg oil

An organoleptic comparison of CO_2 oil of nutmeg (Myristica fragrans) with a steam distilled equivalent reveals much lower α-terpineol and para-cymene levels. These are degradation components caused by steam processing [23]. As a result the CO_2 oil is sweeter, warmer, less harsh and disinfectant pine-like in character. GLC comparisons have been published [22].

Cardamom oil

Organoleptically, the CO_2 extract of cardamom seeds (Elettoria cardamomum) is greener, fresher and more estery in character than steam distilled oil. A GLC comparison [24] shows the CO_2 oil to be higher in contents of geranyl, terpinyl and linalyl acetates as well as citral. Steam distilled oil contains the extra degradation components: para-cymene, γ-terpinene, terpinolene and linalol.

Vanilla absolute

The high natural vanillin and light creamy colour make this CO_2 extract of vanilla beans (Vanilla planifolia) particularly applicable to light coloured cocktail and cream liqueurs [21].

Coriander oil

The CO_2 extract of coriander (Coriandum sativum) has a sweeter, richer tasting, more spicy character than steam distilled oil. Its content of esters is higher and it also contains coriandrin which is not steam distillable. This component and the many other backnote flavours make CO_2 coriander the natural choice for gin and spicy liqueurs, achieving the full rich flavour of the spice.

Summary

Liquid CO_2, used at low temperature, extracts true to nature oils without using heat. Extracts have six clear advantages: no solvent residues or still notes, more topnotes, backnotes, solubility and aromatics. A comprehensive range of 36 CO_2 extracts has been available in commercial quantities since 1982.

CONCLUSION

CO_2 extraction offers opportunities for processing in the alcoholic beverages industry, but mostly for extracted oils for use in compounded flavoured products.

REFERENCES

1. Rose, A.H. (ed) Economic Microbiology Vol. 1: Alcoholic Beverages. Academic Press, London, pp 1-37, 537-689, 1977.
2. Liddle, P.A.P. and de Smedt, P. 8th Essential Oil Congress. Cannes, France, Transcript paper 176, pp 573-8, 1980.
3. Mitchell, C.B. Review of references on extraction with CO_2. Distillers CO_2 Ltd., Cedar House, London Road, Reigate, Surrey, UK, 1982.
4. Tayler, F.M. Review of CO_2 - The solvent for the food and related industries. Wolviston Consultancies, 50 Cricket Lane, Staffs, UK, 1984.
5. Brogle, H. Chemistry and Industry 12, 385-90, 1982.

6. Distillers CO_2 Ltd. Use of CO_2 in Food Processing. Distillers CO_2 Ltd., Cedar House, London Road, Reigate, Surrey, UK, 1987.
7. Kuk, M.S. and Montagna, J.C. Gulf R & D Co., Pittsburgh, PA 15230-101-111, 1983.
8. de Filippi, R.P. and Vivian, J.E. G.B. Patent 2 059 787, 1979.
9. Berger, F., Cerlis, B. and Sagi, F. European Patent 0 077 745, 1981.
10. Wainwright, T. Brewers Guardian **3**, 16-8, 1981.
11. Schultz, W.G. and Randall, J.M. Food Technology **24**, 1282-6, 1970.
12. Stahl, E. and Schilz, W. Chem. Ing. Tech. **48**, 773-8, 1976.
13. Volbrecht, R. Chemistry and Industry **12**, 397, 1982.
14. Calame, J.P. and Steiner, R. Chemistry and Industry **12**, 399, 1982.
15. Weldon, A.G. European Patent 0 061 877, 1981.
16. Fincher, J.M. Unpublished work, 1976-88.
17. Moyler, D.A. and Heath, H.B. 10th Essential Oil Congress. Washington, USA, 1986.
18. Hirata, M. and Ishikawa, T. Supercritical Fluid Technology. Tokyo Metropolitan University, 1987.
19. Gardner, D.S.J. Chemistry and Industry **12**, 402, 1982.
20. Moyler, D.A. Perfumer and Flavorist **9**, 109-14, 1984.
21. Moyler, D.A. and Heath, H.B. Templar Essential Oils, Application to Flavours and Fragrances. Felton Worldwide, Bletchley, England, 1988.
22. Birch, G.G. and Lindley, M.G. (eds) Developments in Food Flavours. Elsevier Applied Science, London, pp 119-29, 1986.
23. Chen, C.C., Kuo, M.C. and Ho, C.T. Journal of Food Science **51**, 1364-5, 1986.
24. Chen, C.C., Kuo, M.C. Wu, C.M. and Ho, C.T. Journal of Agricultural and Food Chemistry **34**, 477-80, 1986.
25. Chen, C.C. and Ho, C.T. Journal of Agricultural and Food Chemistry **36**, 322-28, 1988.
26. Pickett, J.A., Coates, J. and Sharpe, F.R. Chemistry and Industry **7**, 571, 1975.
27. Moyler, D.A. Perfumer and Flavorist, in press, 1988.

29

Summing up: recent developments in the flavour of distilled beverages

A. A. Williams
Sensory Research Laboratories Ltd., 4 High Street,
Nailsea, Bristol, BS19 1BW, UK

The flavours of distilled beverages, whisky and brandy in particular, are the result of a complex and long process. Despite research a successful product still depends very much on the skill of the distiller and blender to create a marketable product. As time goes on, however, we are beginning to understand more about the factors which such people manipulate. Our knowledge of the ingredients of distilled beverages, how they contribute to the flavour of the final product, the mechanics by which they arise and how they can be influenced by the various manufacturing steps, is increasing annually.

At the last symposium on distilled beverages held at Stirling (Piggott, 1983), papers concentrated very much on the volatile aroma constituents of distilled beverages, with a few discussing distillation, sensory analysis and maturation.

The number of volatile components in whisky was at that time around 350 whereas today we can count some 500 flavour components (Philp, 1986). This reflects the increasing skill and ingenuity of the scientist.

Some of the papers at this meeting outlined new information and others how known compounds change during processing, others in fact outlined some of the new techniques which are currently being employed to provide us with more information.

Coupled gas chromatography-gas chromatography was described by H. Maarse (chapter 1). This technique is proving very useful for information on trace components which are so often buried under larger ingredients in single column chromatography. The use of carbon dioxide for extraction of flavours and supercritical fluids in general for separation (M. Raynor, chapter 7) open up new areas and opportunities in the field of extract isolation and separation. Fourier Transform-Infra Red in particular when coupled to a gas chromatograph whether using gas pipe or matrix isolation approaches, is also providing a new insight into the structure of compounds complementing information gleaned by more conventional Gas Chromatography-Mass Spectrometry in its various guises (Electron impact, Chemical ionisation, High and Low Resolution etc.). It just needs Nuclear Magnetic Resonance Spectroscopy to be brought on line to complete the trio.

Although volatiles are important to distilled beverages, as with all alcoholic beverages, they are not the whole story. Flavour is more than just aroma; taste and after-taste are also very important. Various non volatiles, in spirits, particularly those derived from wood extraction, not only gives rise to colour but also to acidity, bitterness and astringency, important characteristics in any product.

Here again research developments in High Performance Liquid Chromatography and Fast Atom Bombardment Mass Spectrometry will eventually enable the scientist to make inroads into the nature and structure of such components and how they influence taste.

Analytical information *per se*, whether it be volatile or non volatile, although invaluable in product authentication (H. Maarse, chapter 1), is however of little use in quality evaluation unless one can determine the sensory aspects they impart and also the influence such characteristics have on the acceptance of a product.

At this meeting we have had a number of papers on sensory techniques, both general and specific (P. Duerr, chapter 3). When relating product perception to acceptance, sensory characteristics are not the only factors of importance; image, presentation, advertising and price all play a role. These along with the sensory properties need to be evaluated and manipulated and optimised if a product is to survive in the market place. M. Baker (chapter 22) discussed some of these non-sensory factors and the problems of marketing, particularly in respect to product development in his presentation.

Laboratory Profiling techniques are obviously a great benefit in characterising how alcoholic beverages are perceived, either in sensory or imagery terms but if product acceptance is to be understood such information needs to be related to consumer likes and dislikes so we can understand which are the important factors from the point of view of acceptance. It also needs to be related to chemical and physical measurements, or even production or marketing factors if we are really to understand and use this information to improve products (Figure 1).

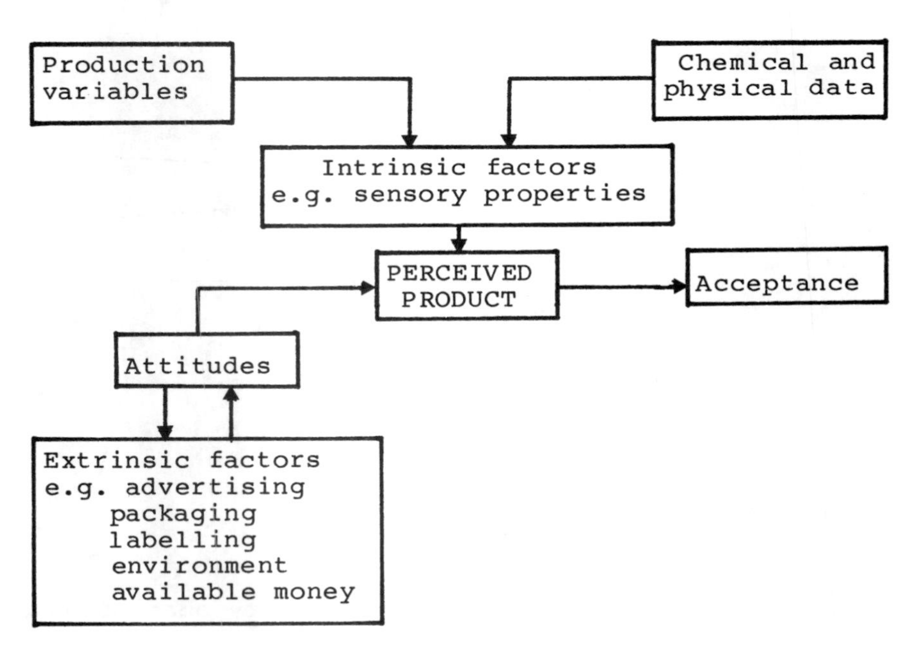

Figure 1. Factors influencing beverage acceptance

At this meeting we have heard about both instrumental and sensory perception of colour and how they relate (D. MacDougall, chapter 8). We have also heard about off flavours and taints (D. Land, chapter 2), when the presence of a small amount of a foreign ingredient can have tremendous influence on the profile of a product and its acceptance.

We have also heard how it is possible to extend descriptive procedures to consumers and obtain information on how they, as opposed to in-house trained panels, perceive products (C. Guy, chapter 4). This can be very important information because if trained panels and consumers perceive products in a different way, relying on trained panels for product development could lead to a manufacturer concentrating on characteristics which the consumer could not care less about, or completely ignoring others which are of paramount importance to the consumer. It is the latter after all who buys the product and on whom a company's success depends, not the trained panel.

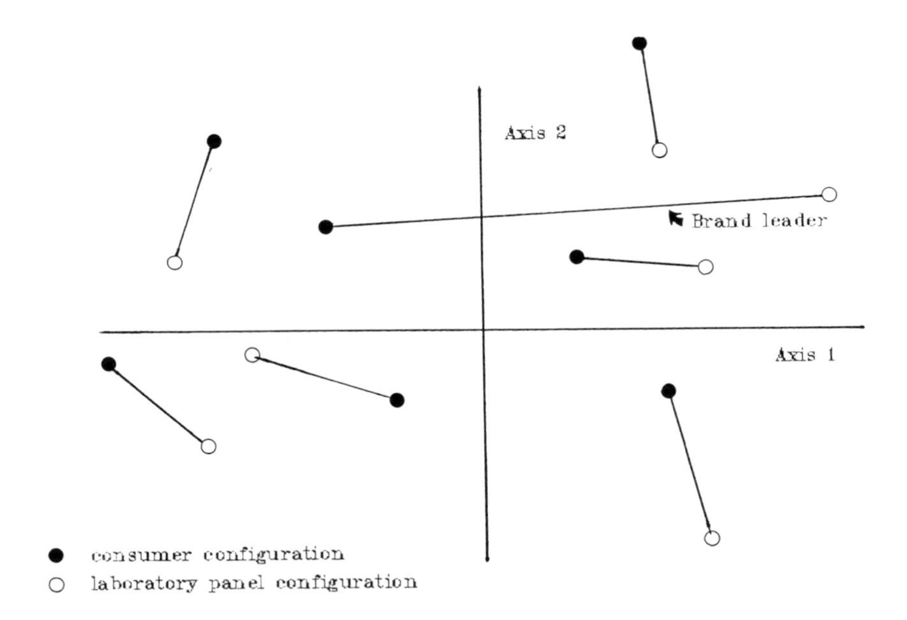

Figure 2. Consumer and laboratory perception of products

Figure 2 provides a sample map where such a difference in perception has occurred. In this case consumers and laboratory panellists perceive most products in the same way. Differences however occur in the perception of the brand leader, the consumer panel perceiving it as having certain characteristics and the laboratory panel as having other characteristics. If Research and Development personnel were to take note of the laboratory panel they could well try to improve characteristics unimportant to their consumers.

Unlike the first symposium in 1983 this meeting began to touch on not just monitoring and measuring factors important to the consumer but also manipulating them.

The various ways in which whisky flavour can be manipulated were described by A. Paterson (chapter 9). As well as altering distillation and blending parameters, selection and genetic manipulation of yeast to produce more desirable flavour ingredients is now becoming a possibility (D.R. Berry, chapter 26). More research of this nature, linking such factors to analytical, sensory and consumer information will really start to put the control of beverage flavour into the hands of the producer. The industry still has a long way to go but trends since the last symposium indicate that science is moving in the right direction. Research of the type described in this book will eventually take the fermented beverage industry to the day when it can truly be said to have control of its products and therefore be in a position to move eventually to complete automation.

REFERENCES

Philp, J.M. (1986). In Proceedings of the 2nd Aviemore Conference on Malting, Brewing and Distilling. Campbell, I. and Priest, F.G. (editors), pp 148-164. Institute of Brewing, London

Piggott, J.R. (1983). Flavour of Distilled Beverages: Origin and Development. Ellis Horwood Ltd., Chichester, U.K.

Index